Understanding Silviculture Technology

This book describes the various technologies used in silviculture with a special focus on newer technologies that have been adopted over the past two decades in plantation forestry practised in the Southern Hemisphere.

Historically, innovations in silviculture technology have been slow to develop; however, over the past two decades, new technologies have been invented and adopted to deal with labour, efficiency, and quality challenges. This book presents a comprehensive introduction to modern silviculture technology, as well as explains the environment and operations within which these technologies are used. In the context of this book, the focus is on industrial plantation forests grown in a range of Southern Hemisphere countries, mainly South Africa and Brazil, with a specific focus on the technologies used in establishing and tending eucalypts, pine, and wattle species. This book examines modern technologies used to produce trees from a nursery, including managing residues on the site, preparing the soil, fertilising, and irrigating the plants. It then examines the technologies available for after planting, including those that can assist in eliminating weeds, pruning, thinning, and fire protection systems. This book also discusses digital technologies available to manage silvicultural operations. By the end of this book, readers will have a strong understanding of the range of technologies available, the impact of those technologies, and their capabilities and limitations.

This book will be of great use to students of silviculture and forestry, as well as professionals working in the forestry industry.

Muedanyi M. Ramantswana is a senior lecturer specialising in Forest Engineering at the Nelson Mandela University, South Africa. He is the founder and main editor of the *Forsilvitech* online newsletter.

Saulo P. S. Guerra is an associate professor at the Universidade Estadual Paulista (UNESP), Brazil, leading the LABB Research Group on Biomass and Bioenergy. He is affiliated with the Institute of Bioenergy Research (IPBEN) and the Nelson Mandela University, South Africa.

José Mauro S. da Silva is a professor at the Federal University of São Carlos, Brazil.

Understanding Silviculture Technology

Plantation Forestry in the Southern Hemisphere

Muedanyi M. Ramantswana, Saulo P. S. Guerra, and José Mauro S. da Silva

Routledge
Taylor & Francis Group
LONDON AND NEW YORK

earthscan
from Routledge

Designed cover image: Muedanyi M. Ramantswana

First published 2026
by Routledge
4 Park Square, Milton Park, Abingdon, Oxon OX14 4RN

and by Routledge
605 Third Avenue, New York, NY 10158

Routledge is an imprint of the Taylor & Francis Group, an informa business

British Library Cataloguing-in-Publication Data
A catalogue record for this book is available from the British Library

ISBN: 978-1-032-48880-6 (hbk)
ISBN: 978-1-032-48654-3 (pbk)
ISBN: 978-1-003-39122-7 (ebk)

DOI: 10.4324/9781003391227

Typeset in Optima
by codeMantra

Dedication

This book is dedicated to all academics, students, and foresters who are passionate about silviculture technology, sustainable forest management, and continuous improvement.

Contents

About the authors

Muedanyi M. Ramantswana holds undergraduate qualifications in Forestry from Nelson Mandela University and a PhD in Forest Science from the University of Pretoria. He is currently a senior lecturer specialising in Forest Engineering at Nelson Mandela University, George, South Africa. Previously, he held various forestry roles, including harvesting, silviculture, and forestry management for Sappi. He is also the founder and main editor of the *Forsilvitech* online newsletter, which disseminates information about the latest global silviculture technologies.

Saulo P. S. Guerra holds a degree in Forest Engineering and a PhD in Agronomy (Energy in Agriculture) from UNESP. He completed his postdoctoral research at the University of Arizona in 2010. Currently, he is an associate professor at UNESP's Faculty of Agronomic Sciences, leading the LABB Research Group on Biomass and Bioenergy, and is affiliated with IPBEN and Nelson Mandela University (South Africa). His expertise includes forest mechanisation and eucalyptus plantations. He served as the director of FEPAF (2010–2012), coordinated UNESP's Forest Engineering undergraduate programme (2012–2018), and led the PCMAF programme at IPEF. He is now the director of UNESP Innovation Agency (AUIN), which ranked first globally in the 2025 *THE Impact Rankings* for SDG 9 (Industry, Innovation, and Infrastructure), and won Brazil's national innovation award for three consecutive years (2023–2025).

José Mauro S. da Silva is a forestry engineer and full professor at the Federal University of São Carlos, Sorocaba campus. Currently working in the LASEM research and extension group (seed and seedling laboratory), he teaches seedling production and forest nurseries at the undergraduate and graduate levels. He coordinates extension projects with several companies in Brazil.

Contributor
Guilherme Oguri
Project and Strategy Manager
IPEF, Brazil

Preface

This book arose from our observation of shifts from manually oriented to intermediate and advanced technology developments in silviculture. Due to our passion and interest in the field of silviculture technology, we recognised the need for a teaching resource that students and academics could use to learn about the various technologies available in Brazil and South Africa. Through our country-specific silviculture technology surveys, we recognised that there is a knowledge gap in understanding the applications of various silviculture technologies available in plant production, regeneration, and stand tending. Furthermore, we recognised that there are similarities and differences in how forest practices are applied in both countries, and we wanted to expose these aspects through this book. Our aim with this book is to practically present various current and possible future technologies in silviculture, while revealing the challenges and opportunities for integrated forest value chain activities. Through active collaboration and interaction between Nelson Mandela University, UNESP, and IPEF, we decided to compile this textbook, which is arguably the first of its kind in this field. We hope this book will be a valuable learning resource for all who peruse its pages and glance at the figures presented. We would like to thank everyone who contributed to making this book possible.

Muedanyi M. Ramantswana
In George, South Africa,
on 1 September 2025

Preface

[illegible]

Acknowledgements

The authors would like to thank the following individuals and organisations who enabled this book to be compiled:

Muedanyi M. Ramantswana:

- I want to express my deepest gratitude to my family (Nontethelelo, Muandisi, and Muhuliswa), my father (Jack Nndwakhulu Ramantswana), and many close family and friends (work and church) for the continuous encouragement along the journey.
- I would like to thank my co-authors and contributors for enabling this book to transform from ideas to reality.
- I would also like to thank Nelson Mandela University for allowing me to compile this book and, importantly, all my colleagues in the Forestry department.
- To the Nelson Mandela University undergraduates and postgraduate students who have been an encouragement during this journey, I hope this book will be a valuable tool in your education and future careers.
- To the various industry professionals in South Africa and globally, whom I had the pleasure of interviewing and discussing different aspects of this book, thank you.

Saulo P. S. Guerra:

I want to express my deepest gratitude to my beloved wife, **Isabel**, and my daughter, **Isabella**, for their unwavering support, patience, and encouragement throughout this journey. Their love has been a constant source of strength.

I am also grateful to the São Paulo State University (UNESP) for providing the academic environment that made this work possible. Special thanks go to my colleagues and mentors who have contributed to my growth and inspired me along the way.

I extend my sincere appreciation to the IPEF (Institute of Forest Research and Studies), PCMAF (Cooperative Program on Forest Mechanisation and Automation), and PPPIB (Cooperative Program on Potential Pine Productivity in Brazil), which have served as vital links between academia and the forest-based industry. Their collaboration and dedication to sustainable forestry have enhanced this work and increased its relevance to real-world applications.

To all those who have supported me – family, friends, and professionals in the forestry sector, thank you for being part of this journey.

José Mauro S. da Silva:

I would like to extend my gratitude to my co-authors, Dr Muedanyi Ramantswana and Prof Saulo P. S. Guerra, for the opportunity to contribute to this book.

1 Introduction

Muedanyi M. Ramantswana, Saulo P. S. Guerra, and Guilherme Oguri

1.1 Overview and objectives of the textbook

This book aims to comprehensively describe the various technologies used in silviculture with a special focus on newer technologies that have been adopted over the past two decades in plantation forestry practised in the Southern Hemisphere. Furthermore, this textbook will contain simple illustrations to improve the readability and understanding of the content by all forestry stakeholders interested in the subject matter. This book primarily focuses on re-establishment, tending, and, to a lesser extent, harvesting technologies used in South Africa and Brazil. To grow and produce wood sustainably from a forest requires good silviculture to be practised. Good silviculture can be practised by utilising technology to enhance efficiencies, safety, quality, and minimise environmental impacts. This book is structured to follow the sequence of the forestry value chain. Chapter 1 provides an overview of plantation forestry practices, technology, and planning involved in the cultivation and management of a forest. Chapter 2 focuses on nurseries, describing the resources and technologies used in plant production. Chapter 3 discusses the methods and innovations employed in regenerating/re-establishing forests, focusing on activities such as residue management, preparing planting positions, weeding before planting, tree planting, and fertilisation. Chapter 4 describes stand-tending activities and the latest tools and resources available for executing activities such as weeding after planting, forest protection, pruning, and thinning. Chapter 5 reflects on past, emerging, and future trends in the adoption of silviculture technology. Chapter 6 focuses on the relationship between silviculture and harvesting operations, specifically the impacts of establishment and tending activities on harvesting operations, as well as the reciprocal effects of harvesting operations on these activities. Chapter 7 brings together the key learnings from all chapters and reflects on prospects for technology development in silviculture.

1.2 Global plantation forestry

According to the United Nations (2023), the global population reached 8 billion in 2022, and it is projected to grow to 8.6 billion by 2030 and

DOI: 10.4324/9781003391227-1

9.8 billion by 2050. The population increase is mainly due to more people reaching reproductive age, a gradual rise in life expectancy, urbanisation, and increasing migration. As the population grows, economies expand, and the demand for wood – used for energy, heating, cooking, building, furniture, and paper – also rises. Globally, in 2021, 770 million people lacked access to electricity, and over 2.5 billion lacked access to modern cooking facilities (REN, 2022). With this rising demand and need for forest resources, the sustainable regeneration and management of productive forests is essential to meet the current and future demands. Forests cover approximately 4.06 billion hectares worldwide, spanning boreal, temperate, subtropical, and tropical zones (FAO, 2022). Approximately 34% of these forests comprise native tree species with minimal human disturbance. Plantation forests are those established by humans using seeds, seedlings, or clones at specific spacing (West, 2014). In 2020, planted forests covered 294 million hectares, constituting 7% of the global forest area. These forests are created through planting or deliberate seeding (FAO, 2020a). Plantation forests account for 3% (131 million hectares) of the total forest area worldwide and are defined as forests managed intensively, usually comprising one or two species, even age classes, and regular spacing (FAO, 2020a), mainly grown for industrial wood products. In South America and Southern Africa, plantation forests include non-native species, whereas in North and Central America, they consist of native species (FAO, 2020b).

Industrial plantation forests grow faster than natural ones, providing wood resources to meet the growing global population and easing pressure on natural forests. In 2020, the industrial roundwood supply nearly reached 2 billion m^3, with 44% coming from naturally regenerated and boreal forests, while planted forests supplied 46% of the demand. Intensively managed plantation forests produced six to eight times more industrial roundwood than the global average (FAO, 2020a). According to FAO (2020a), 33 million hectares of highly productive plantation forests have the potential to meet the increased demand for industrial roundwood up to 2050, which is estimated to reach 3.1 billion m^3 by then.

1.3 Plantation forest compositions

The global plantation forest area of 131 million hectares comprises native (62.1%) and introduced (37.9%) species (FAO, 2020a). The two regions with the largest plantation areas featuring introduced species are Asia (20.9 million hectares) and South America (17.8 million hectares) (Table 1.1). Plantation forestry can involve mixed species or monocultures focused on cultivating a single species with desirable traits (e.g. rapid growth, good wood properties and disease resistance) (Sands, 2013). The most widely planted monoculture genera globally are *Eucalyptus*, *Pinus*, *Acacia*, and *Tectona* (Kelty, 2006). These species are cultivated for 10–15 years for pulp and paper, and for 20–30 years for sawtimber production (West, 2014). Monocultures benefit

Table 1.1 Plantation forest area by region (FAO, 2020a)

Region	*Plantation forest area (ha)*
Africa	7,678,000
Asia	79,332,000
Europe	4,495,000
North and Central America	15,177,000
Oceania	4,356,000
South America	20,099,000
Total	131,137,000

from simplified stand management practices (Kelty, 2006), including activities such as nursery production, stand establishment, and harvesting. Despite the many advantages of plantation monocultures, several issues are associated with them, such as biological risks, instability of trees on exposed soils, poor aesthetics, and a reduction in conservation value and biodiversity (Savill et al., 1997, Stephens and Wagner, 2007, Freer-Smith et al., 2019). Several other plantation species are being expanded for the cultivation of palm oil, rubber, and bamboo (Liu et al., 2018). However, this book focuses on plantation forestry monoculture practices in the Southern Hemisphere, particularly in Southern Africa (South Africa) and South America (Brazil), with the focus on *Eucalyptus*, *Pinus*, and *Acacia* genera.

1.3.1 South African forestry industry

The total land area of South Africa is 122.3 million hectares, of which 1% is commercial plantation forestry (Oberholzer, 2021). The plantation forest area in South Africa covers 1,194,663 million hectares, with limited expansion opportunities (Godsmark and Oberholzer, 2020). The most widely planted commercial genera in South Africa are *Pinus* (48.6%), *Eucalyptus* (44.2%), and *Acacia* (6.8%), with other species (0.3%) (Oberholzer, 2021). The plantation forest area is distributed across five provinces (Mpumalanga, 41.2%; KwaZulu-Natal, 40%; Eastern Cape, 11.9%; Limpopo, 4.3% and Western Cape, 2.6%). It is anticipated that new afforestation of 151,000 ha of productive plantations will be established in the Eastern Cape, Limpopo, Mpumalanga, and KwaZulu-Natal within the coming years (Strategy Execution Advisers, 2020). Apart from this restricted expansion, no large land masses are available to increase wood production. Most of the existing forested land is owned by corporate companies (51.3%) and commercial farmers (14.6%). Therefore, to stay competitive, South African organisations must improve efficiency and reduce costs by producing timber at the lowest possible price (Zwolinski and Bayley, 2001).

South Africa re-establishes an estimated 72,000 ha annually, of which 75% is replanted and 25% is regenerated using coppice methods (Oberholzer, 2023). Furthermore, in 2023, the South African forestry industry produced

13.37 million tonnes for the year, comprising pulpwood (68.7%), sawlogs (26.3%), poles (2.1%), mining timber (1.7%), and other products (biofuel) (1.2%) (Forestry South Africa, 2023). The forestry industry plays a vital role in the economy by providing direct and indirect employment opportunities to over 147,000 people, who have an estimated 500,200 dependents. Overall, the gross value of output from forestry and forest products is estimated to be R36.3 billion (Oberholzer, 2021).

In South Africa, only 70% of the forest land is planted. The remaining areas comprise conserved riparian zones, grasslands, and indigenous forests, which provide an essential habitat for animals and unique endemic species. The forestry industry also supports biodiversity and conservation of rare species within the broader plantation forest areas (Pott, 1997). Furthermore, there are strict environmental guidelines which are adhered to when performing various forestry activities to mitigate and minimise negative impacts to the surrounding environment (FSA Environmental Management Committee, 2019). Most of the wood produced from plantation forests in South Africa is certified under various certification bodies to guarantee sustainable forest practices (Scotcher, 2006).

Silviculture activities in South Africa are still manually oriented; however, over the past 15 years (reference year 2025), there has been an increased adoption of new technologies to improve safety, productivity and labour-related challenges (e.g. increased cost). The adoption and utilisation of technology are vital for competitiveness and sustainability (Ramantswana et al., 2021). Technology serves as a tool within the silviculturist's resources to facilitate the implementation of effective silvicultural practices that enhance the survival, growth, and uniformity of trees.

1.3.2 Brazilian forestry industry

The Brazilian forest sector, particularly concerning eucalyptus plantations, is characterised by various practices to ensure productivity and sustainability. *Eucalyptus* is a dominant genus in Brazil's planted forests, covering 7.8 million hectares and accounting for 76% of the total planted area, underscoring its significance to the sector. The sector utilises fast-growing, non-native species like eucalyptus due to their short rotation cycles and high profitability. Planting methods vary from manual to fully mechanised, with a growing preference for mechanisation to address labour shortages and reduce costs. Additionally, it reduces labour stress and the risk of accidents or injuries. Brazil has approximately 10 million hectares of planted forests, and the production area continues to increase. The National Plan for the Development of Planted Forests (PNDF) highlights the significant potential for expansion, especially in degraded areas and underutilised pastures. Additionally, Brazil has committed to restoring 12 million hectares of forests by 2030 as part of its climate goals under the Paris Agreement. This effort involves collaboration from various sectors, including government, NGOs, and the private sector, to achieve large-scale restoration (Brasil Ministério da Agricultura e Pecuária, 2024).

In Brazil, pine plantations cover a significant area, although they are less extensive than those of eucalyptus. According to recent data, pine forests cover approximately 1.6 million hectares. These plantations are primarily located in the southern states of Paraná, Rio Grande do Sul and Santa Catarina, the country's leading producers of pine. Pine is used for various industrial purposes, including the production of wood panels, laminated floors, pulp, paper, and biomass. The sector has shifted towards more sustainable practices and technological advancements to improve productivity and efficiency (Brazilian Tree Industry, 2024).

Sustainability and conservation are key priorities for the sector. The industry conserves a significant area, totalling 6.91 million hectares, including 4.88 million hectares of Legal Reserve and 1.94 million hectares of Permanent Preservation Areas. The sector is committed to biodiversity conservation and actively develops biodiversity strategies. It adopts integrated crop–livestock–forest systems (ICLF) that combine sustainable food, fibre, and energy production while reducing greenhouse gas emissions. The use of high-density plantations is being studied for its effects on dry matter yield, energy yield, and nutrient exports, with higher planting densities found to increase biomass yield (Brazilian Tree Industry, 2024).

The economic and social impact of the forest sector is significant. The industry plays a key financial role in Brazil, contributing to employment and tax revenues. The Brazilian planted tree industry generates R$11.4 billion in taxes, representing 0.9% of the country's total tax revenue. The sector has an international trade volume of US$12.7 billion and is a major global exporter of cellulose. The industry emphasises partnerships with small- and medium-sized producers and promotes economic diversity in local communities. So, forest programmes are essential, where companies support the growth of planted forests for industrial purposes (Brazilian Tree Industry, 2024).

The industry is focused on technological advancement. It invests in research and technology to improve productivity and efficiency and has registered 1,279 patents or intellectual properties. Advanced technologies, such as remote sensing and data analysis, enable the mapping and monitoring of planted areas. The industry is also increasingly adopting mechanised operations, particularly planting, to reduce costs and address labour shortages. The industry's production and market aspects are robust. It has a high production capacity and is a leading producer of cellulose, with a significant amount of its production exported. In 2023, Brazil produced 24.3 million tonnes of cellulose. The sector uses various products, including timber, pulp, paper, wood panels, and laminates (Brazilian Tree Industry, 2023).

The Brazilian forest sector is committed to balancing economic growth with environmental and social responsibility. The focus on sustainable practices, technological innovation, and biodiversity conservation ensures that the industry can continue to thrive while protecting the environment and supporting local communities.

1.4 Silviculture in plantation forests

Silviculture is the art and science of cultivating forest crops through managing the establishment, growth, composition, health, and quality of forests and woodlands to meet the diverse needs of landowners and society sustainably (Savill et al., 1997). Silviculture includes silvicultural regimes, which are planned programmes of silvicultural operations carried out during the life of a stand of trees. These encompass establishing or regenerating the stand and tending it during the rotation to produce specific products (Shepherd, 1986).

However, the definition of silviculture has been reshaped due to current challenges, mainly influenced by climate change (Zhang et al., 2018, Szmyt and Dering, 2024) and new social demands (Branca et al., 2020). Along with these definitions, new techniques and technologies applied to forest management are emerging, adapting to new realities and trends. These topics will be addressed in the following chapters of this book (da Silva, 2023).

The interpretation of foresters regarding the best management practices to be used is closely related to local edaphoclimatic conditions, culture, available technologies, and the knowledge acquired in the past (Achim et al., 2022). However, it is essential to report on the most used silvicultural systems. A study involving 21 mostly Canadian institutions reported a cultural shift in silviculture from an empirical approach to a new concept of "silviculture as a scientific discipline" (Achim et al., 2022). This latest proposal consists of three key themes: observation, anticipation, and adaptation (Achim et al., 2022).

1.5 Understanding silvicultural systems

A silviculture system is a planned programme of silvicultural treatments throughout a stand's entire life, involving cutting, reproduction, tending operations, and intermediate cuttings (Matthews, 1991). Evans and Turnbull (2004) described a silvicultural system as a method for growing and regenerating trees. According to the literature, Nyland (2016) summarised findings that show all silviculture systems comprise three main treatment components or functions: regeneration, tending, and harvest. Any silvicultural system is cyclic, with the various components being interdependent and ongoing (Figure 1.1).

Nyland (2016) states that regeneration of a stand may be achieved naturally and artificially (seeding and planting). Once a stand has been established, the trees may receive tending treatments (pruning and thinning) to encourage subsequent growth and development. Harvesting is used to clear and enable regeneration of the stand when the trees have reached a specific age class, depending on the management objectives. Even though silviculture does refer to regeneration, stand tending, and harvesting, we have practically categorised the processes into silviculture (mainly referring to regeneration

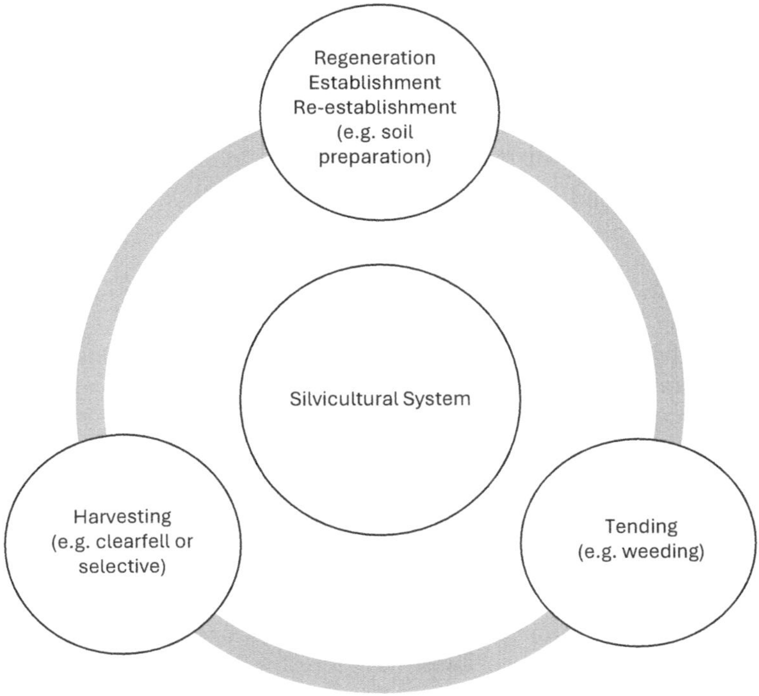

Figure 1.1 A silvicultural system, which is a continuous cycle, and the interdependence of its components

and tending activities) and (ii) harvesting (clear-felling, extraction, and transport). Hence, within the context of this book, the term silviculture is used with a focus on the regeneration and tending process stages. The final chapter describes the relationship between "silviculture and harvesting" to identify the interdependencies and impacts of the silviculture stages (regeneration and tending) on harvesting stages and vice versa. Ultimately, a combination of the sequence and kind of treatments determines the silvicultural system forms and creates the character and development of the resultant forest. Any silvicultural system must be aligned with the landowner's objectives (Nyland, 2016). The definition of these objectives may seem straightforward; however, this task becomes more complex when various stakeholders are involved in management decisions (Puettmann et al., 2015). External stakeholders, such as communities near the forests, influence forest production objectives and the management approaches adopted in these forests.

Two broad categories classify silviculture systems: intensive silvicultural and less intensive silvicultural systems. An intensive silvicultural system comprises several operational applications conducted sequentially with high input costs through the life of a stand. In contrast, a less intensive silvicultural system involves a less structured management approach with low input costs

(Smith, 1986). The silviculture applied in plantation forests and natural forests may differ depending on the forests' structure, dynamics, and functions (Gamborg and Larsen, 2003).

1.6 Objectives of silvicultural systems

According to Nyland (2016), the goals of a sound silviculture system should include the following:

- Optimise the benefits of the stand by capitalising on the full potential of the site
- Shorten the investment period by manipulating silvicultural factors to avoid delay in bringing the required condition
- Contain costs and optimise the sought-after values
- Sustain ecosystem health and productivity by applying appropriate practices that are ecologically and biologically friendly
- Provide a stand of trees and vegetation suited to the landowner's needs

According to Shepherd (1986), hundreds of silvicultural system routines exist globally; some are more complex than others; however, the basic needs need to be clearly defined. Silviculture systems can be classified as high forest methods and coppice forest (low forest) methods. Each system comprises treatments that include a range of reproduction methods, for example, clear-felling, seed tree, shelterwood, selection, and two-stage methods (Nyland, 2016), (Table 1.2).

Table 1.2 Silvicultural systems (Smith, 1986)

Forest	*Stand*	*Method*	*Description*
High forest	Even aged	Clear-cutting	One clear-felling of the entire stand and reproduction through natural seeding or artificial regeneration
		Seed tree	One clear-fell of old stand, except for a small number of seed trees
		Shelterwood	Removal of the old stand through a series of cuttings over the rotation. Preproduction under the shelter of seed trees
	Uneven aged	Selection	Remove mature timber in small groups or scattered individuals at short intervals, repeated indefinitely
Coppice forest		Coppice	Cutting in which dependence is placed on vegetative regeneration
		Coppice with standards	Production of coppice and high forest in the same area. Trees with seedling origin carried through much longer

1.6.1 Plantation forest – even-aged clear-cutting silvicultural system

The clear-fell system involves managing successive even-aged stands by clear-felling the entire stand of trees at planned rotations and then re-establishing and tending a new stand in place of the original one (Smith, 1986, Shepherd, 1986). According to Evans and Turnbull (2004), most plantations are established by planting rather than direct sowing, and at maturity, they are clear-felled and replanted rather than naturally regenerated, unless regeneration occurs from coppice shoots. Replanting is a preferred re-establishment method because of increased mechanised felling, insufficient stand stocking due to damaged coppice stumps, risk (biotic or abiotic) and, to a lesser extent, tree improvement work because it decreases with each generation. In plantation forestry, the even-aged clear-fell silvicultural system is preferred as it is a relatively simple, cost-effective, and reliable method for achieving uniform regeneration compared to other alternatives, such as natural regeneration. Uniformity is essential, especially when dealing with fast-growing species in plantation forestry (Evans and Turnbull, 2004). Commercial plantation forest operations comprise silvicultural practices such as planning, site preparation, species/clone selection, optimal spacing/density/ planting, weed management, irrigation management, nutrient management, pest and disease management, and harvest management (Seenivasan and Parthiban, 2017). This book will focus on technologies applied in the even-aged clear-fell silvicultural system. Even though other silvicultural systems described in Table 1.1 are important in the global forestry context, this book will focus exclusively on regeneration through establishment, re-establishment, and coppice methods.

1.7 Silviculture technology progression

Forest technology is "the study of the operations, work methods, logistics and economics related to forest resources utilisation" (Uusitalo and Pearson, 2010). Silviculture technology, in the context of this book, can be described as the application of scientific knowledge to study modern tools, machinery, and equipment for establishing, tending, and growing trees. Globally, technology is advancing rapidly across various industries, driven by the development of digital (e.g. the Internet of Things, artificial intelligence), physical (e.g. autonomous vehicles), and biological (e.g. genetic engineering) technologies (Li et al., 2017). According to Ramantswana et al. (2021), the drivers of modernisation in silviculture are:

- Improvements in health and safety
- Increased operational productivity and reduced costs of operation
- Improve stand productivity (quality)
- Social aspects/issues (mainly labour related)
- Environmental aspects/issues
- Other aspects/issues, such as climate change and global competition

To understand the current status and use of technology, it is necessary to reflect on the progression of technology from the past to the present (Table 1.3). Despite the increased use of technology globally, silviculture operations remain largely manually oriented. In 2018, it was estimated that 85% of all silviculture operations were performed manually (Choudhry and O'Kelly, 2018).

Table 1.3 Summary of silviculture technology progression in South Africa and Brazil

Country	*South Africa*	*Brazil*
Activity	*Site preparation*	
Before 1990	• Residue management by burning • Soil preparation – ploughing, disking, rotovating, ridging, ripping, and subsoiling	• Residue management – burning was routinely practised • Soil preparation – harrowers
1990–2010	• Residue management by burning is still dominant • Soil preparation – from 2000s, focus shifted to cost-effective pitting methods and implements	• Residue management – burning not permitted • Soil preparation – subsoilers were adopted
2010–Current	• Residue management by burning and the introduction of mulching and biomass collection • Soil preparation – earth augers and pitting machines were tested, proving effective alternatives to manual pitting	• Residue management – harvesting residues are now moved sideways to enable a clear ripping line. Subsoiling with a V-shear blade helps to reduce stumps and re-align forest plantation rows • Soil preparation – subsoiling, pit preparation on steep areas, and excavator-mounted subsoiler
Activity	*Tree planting*	
Before 1990	• Manual using planting hoes and trowels	• Manual planting methods
1990–2010	• Manual planting using planting hoes and trowels	• Planting tubes • Continuously advancing machines for planting • Semi-mechanised planting – tractor coupled with planting systems
2010–Current	• From 2010, planting tubes and tractor-based semi-mechanised planting systems were introduced • 2025 fully mechanised excavator-based planting machines introduced	• Mechanised excavator-based single-head planting machine or double-head planting machine • Triple-head planter • Forwarder-based planting machines – auto pilot | GPS embedded

(*Continued*)

Table 1.3 (Continued)

Country	*South Africa*	*Brazil*
Activity	*Stand tending*	
Before 1990	• Weeding conducted manually through slashing • Pruning performed manually using pole saws	• Manual weeding techniques • Introduction of fertilisation
1990–2010	• Weeding shifted to herbicides, replacing manual methods • Fertilising conducted using granular fertiliser • Pruning remained manual	• Slow-release fertilisers introduced • Incorporation of fertilisers with subsoiling operations
2010–Current	• Weeding performed using semi- and fully mechanised equipment, such as tractor-mounted boom sprayers, tractor-mounted wind boxes, tractor-disking, and drones • Fertilise with granular fertiliser and tablets • Pruning – motorised pruning introduced	• Weeding – use of pre-emergent and post-emergent herbicides • Weeding performed using semi- and fully mechanised equipment, such as tractor-mounted boom sprayers, tractor-mounted wind boxes, and drones

This reflects a similar situation to the Brazilian market, where 84% of forestry plantations involve manual planting activities (Kotovic Filho, 2016), with only 16% of the remaining areas being semi-mechanised or mechanised.

According to Guerra et al. (2023), the level of mechanisation for overall and leading silviculture operations is presented for each specific area, where the implements or assemblies at each level are used. The following definitions have been adapted from (Guerra et al., 2023):

Manual operation: any activity using tools that are exclusively human-powered, e.g. a pick for pitting, a slasher for weeding, a planting tube for planting.

Semi-mechanised operation (includes motor manual) operation: any activity that involves human labour with the assistance of a machine to perform silviculture operations. A motor can drive the tools or equipment used in these operations, or they can comprise mechanical assemblies with implements that require an operator and at least one or more external person(s), e.g. an auger for pitting, a motorised pruning saw for pruning or a continuous planter or tractor boom spray with lances.

Mechanised operation: any activity using mechanical assemblies or implements that are exclusively mechanically powered with or without autopilot functions, e.g. single pitting head machine, tractor-drawn subsoiler, specialised subsoiler drawn by a tractor with geo-technology systems. Even though the primary work is conducted mechanically, these operations may require manual or mechanical replenishment of resources (e.g. seedlings, water, and fertiliser).

Automatic operation: any activity using tools controlled by various artificial systems, with a high degree of technology and low levels of human intervention. The difference between automated and autonomous operations lies in the degree of human influence involved: autonomous operations involve a high degree of independence and artificial intelligence, such as a drone for standing assessment or an autonomous planting machine. Even though the primary work is conducted partially or completely autonomously, these operations may require manual or mechanical replenishment of resources (e.g. seedlings, water, and fertiliser) needed for a specific operation.

1.8 Silviculture technology adoption challenges

Technology adoption is how people, businesses, or institutions accept and apply new technologies by learning and becoming familiar with the innovation. Similar to agriculture, productivity in forestry can be improved by introducing forest technologies and management systems (Doss, 2006). However, some silvicultural technologies have encountered several challenges or barriers. Some of these challenges hinder progress, and in some instances, the need to resolve the obstacles accelerates technology adoption. The drivers of technology are described in this section, exploring the barriers to the adoption of silviculture technology.

1.8.1 Availability of technology

For many years, the progressive developments of certain silviculture technologies (e.g. planting and weeding) have lagged. This can be attributed to the low costs of manual operations compared to mechanised operations. Due to the lack of available technology to perform some silviculture operations, manually oriented methods have dominated the industry. The development and maturity of technology take time and considerable financial investments by different stakeholders, driven by various factors, such as the need to reduce costs and improve productivity and quality of operations.

If we analyse the introduction of mechanised equipment in harvesting, we notice some characteristics that impact the availability of technology in silviculture:

- Similarity of activities: It can be said that the harvesting process is quite similar regardless of the country, meaning that harvesting a tree in Brazil is very similar to harvesting in a Scandinavian country. This fact has facilitated the import/export of mechanised systems in countries with little tradition

in developing forestry equipment. However, silvicultural processes (e.g. planting, post-planting irrigation, among others) vary by country, mainly in the type and intensity of management and species used.
- Volume of machine production: This factor is related to the previous item, as the same harvesting machine can be used in different countries, with some minor adaptations when necessary. This increases the production volume compared to equipment for specific uses, as in silviculture.
- Hazardous nature of the activity: Harvesting activities present high risks to human health; thus, naturally, the sense of urgency and investment have been greater than those of silvicultural activities.

1.8.2 Lack of investment

Until recently, very little investment had been made in improving silviculture technology. When compared to harvesting, silviculture has lagged. Many more investments in technology have been dedicated to mechanising harvesting operations. Some of the reasons for this approach are that, from a risk perspective, harvesting operations pose a significantly higher safety risk than silviculture and are labour-intensive. This scenario is changing as more investments are being made into developing new silviculture technologies to improve productivity, health and safety, and quality and reduce environmental impacts.

1.8.3 Complexity of certain silvicultural activities

Silvicultural operations are diverse, complex, and influenced by several environmental factors. Therefore, developing technology to deal with complex operations is challenging and costly. The less complex operations (e.g. soil preparation) have been modernised for some time. However, more complex operations like planting have taken some time. Planting is complex because fragile plants must be handled carefully and planted in suitable conditions to ensure their survival and growth. Improving nursery plants (e.g. paper pots) and planting technology are contributing to resolving this complexity.

1.8.4 Physical-environmental barriers to technology adoption

The terrain aspects, such as soil condition, ground roughness (rocks), and slope, are major physical factors impeding modernised operations during re-establishment (Ramantswana et al., 2021). Furthermore, stumps on the site can prevent the effective use of ground-based silviculture equipment when re-establishing a stand.

1.8.5 Socio-economic barriers

Regarding socio-economic barriers, South Africa has a high unemployment rate – 32.9%, compared to the Brazilian rate of 6.2% (Statistics South

Africa, 2019, Estatística, 2025). Therefore, socio-economic barriers – such as the perceived negative impacts of modernisation on employment in rural communities – are essential to consider within the South African and Brazilian forestry context. When mechanising, it is important to balance the social benefits of manually oriented operations (employment) against the economic benefits of using efficient mechanised alternatives (Chapman and Allan, 1978). Mechanisation options can be introduced effectively in areas where labour is unavailable, a shortage of skilled labour in remote regions, low attractiveness of manual labour for younger generations or the need to increase productivity and reduce operational costs (Soler, 2020).

1.8.6 Psychological factors

Looking at the challenges of adopting new technologies beyond those directly related to silvicultural factors, it is important to report the psychological factors that can indirectly interfere with using the latest technologies. These psychological factors can be related to both early adopters and end-users, such as concerns and fears about possible productivity loss and changes in the accustomed way of working (Kratzer et al., 2017).

A literature review that addressed this topic cited the main factors as risk aversion (personality factors), trust between the technology provider and user (attitude factor), personal incentive and fear of technology failure (motivation factors), perceived risks associated with introducing a technology (cognitive factor), validation seek from peer networks (social factors), and organisational culture (organisational factor) (Roberts et al., 2021, Soler, 2020).

1.9 Planning silviculture operations

Silviculture planning can be described as a process of custom designing and optimising the density of regeneration, the timing and intensity of stand-tending treatments and the timing of future harvests (OMNRF, 2015). Concerning technology, this process would involve the identification of tools and equipment which can be used to achieve the treatment objectives. Identifying the technologies required to establish, plant, and tend a forest stand depends on various factors in and around the area. For manually oriented operations, the central resource utilised is labour; hence, planning all operations based on general labour requirements to complete the various tasks is essential. When equipment is introduced to complement labour resources, the number of general labourers required is reduced. For example, if all things are equal, the labour requirements for manually pitting the equivalent area are reduced if a pitting machine is introduced.

1.9.1 Strategic plan

Strategic planning establishes broad goals and land uses. In silvicultural operations, strategic planning considers information about the land (productivity and yield potential), selection of species/clones, and stand density (Seenivasan and Parthiban, 2017, Rubilar et al., 2018). The strategic plan comprises long-term plans that involve identifying different silviculture treatments that can be applied, the cost of executing those treatments, and the expected future results. These have to be aligned with forecasted market requirements to meet the future timber demands.

1.9.2 Tactical plan

Tactical planning describes how strategic goals can be achieved through the use of decision-support tools that predict the consequences of specific treatments on future outcomes. This period is called medium-term planning and is considered a time range of three to five years. During this period, grower companies and contractors can evaluate and align the harvesting plans to the silvicultural plan. Understanding whether the area is being established or re-established is essential because the technology requirements may differ.

Essential factors to consider when planning for technological use in silviculture operations within the medium term:

a Machine utilisation

The capital cost of purchasing equipment is high and must be recouped over the machine's useful life. Before introducing technology into silviculture operations, the machine's planned work must be quantified to ensure that it can be used at acceptable levels to recover the costs over its useful life. Machine utilisation can be derived from the machine's productivity and the area available to work over its lifetime.

b Terrain conditions

The terrain conditions in which technology can be used directly influence the utilisation of the machine. Terrain classification is the grouping of forest land according to the accessibility of an area for forest operations (Erasmus, 1994). Terrain refers to ground conditions (soil-bearing capacity affected by moisture), ground roughness (obstacles like rocks) and slope (inclination and gradient) (Table 1.4). All these factors influence the technologies that can be applied to specific areas. For example, wheeled

Table 1.4 Classes of terrain classification components (Erasmus, 1994)

Ground conditions	*Ground roughness*	*Slope (%)*	*Slope category*	*Slope forms*
Very good	Smooth	≤11	Level	Regular, terraced, undulating, concave, convex
Good	Slightly uneven	>11 to ≤20	Gentle	
Moderate	Uneven	>20 to ≤30	Moderate	
Poor	Rough	>30 to ≤35	Steep 1	
Very poor	Very rough	>35 to ≤40	Steep 2	
		>40 to ≤50	Steep 3	
		>50	Very steep	

machines are typically able to overcome severe ground roughness (stones, rocky conditions) compared to tracked machines. Tracked machines can tackle steeper conditions compared to wheeled machines.

c Social economic conditions

The prevailing socio-economic condition in a specific area needs to be considered before technology is introduced. In countries like South Africa and Brazil, plantation forestry is practised in rural areas with few employment opportunities. One important socio-economic factor is labour availability within or surrounding the forestry area. If labour is available and willing to perform silvicultural tasks, preference must be given to giving them work opportunities before investing in technology. Employment opportunities can help support rural and impoverished communities in the area.

d Health and safety

The health and safety risks of existing methods must be considered when adopting modern technologies in silviculture over the medium term. Tasks with the highest risks (potentially causing harm) to manual workers must be prioritised. For example, functions with poor ergonomic postures that can lead to severe musculoskeletal problems when performed over time must be modernised first if the technology is available.

e Technology backup support

Before modernising silviculture operations, the backup and support available after the technology purchase must be considered. Modern tools and equipment can have mechanical failures; therefore, factors such as the availability and cost of spares, the manufacturer's warranty, the dealer support network, and the turnaround time for repairs and maintenance must be considered.

f Training requirements

The technology's training requirements need to be well understood. Some training must be prescribed by country-specific legislation. The training duration and costs, retraining intervals, and accreditation processes required before and after using the technology must be considered in their entirety. A training plan and schedule must be developed before the equipment is adopted into operations.

g Integration to existing methods

Silvicultural activities are interlinked and can have a direct influence on each other. For example, treating residues can influence subsequent operations such as soil preparation and planting. When technology is introduced, its impact on downstream activities must be assessed. For example, mulching can be conducted to reduce and redistribute residues instead of burning residues after harvesting. However, one must consider that when a stand is mulched, remarking of the stand will be required, compared to a burnt stand, where existing stumps can be used to identify the planting positions for the new rotation.

1.9.3 Annual planning

Annual planning of modern silvicultural operations is typically based on a 12-month cycle. The detailed budget for each activity is determined by considering labour requirements, equipment, and resource requirements (e.g. chemical, fertiliser, pesticides, among others). The areas to be harvested, established, re-established, and tended are derived for planning purposes. The harvesting annual plan of operations (APO) must be integrated into the silviculture plan, as they depend on each other. If the climate conditions limit the performance of specific activities, then activities need to be scheduled according to optimal conditions. For example, in South Africa (in-land areas), re-establishment (residue burning – not allowed in Brazil, soil preparation, and planting) are limited to specific periods (summer periods) and not dry periods because of climatic conditions.

Established stands requiring tending activities must be scheduled throughout the year to ensure they are done optimally (timing). For example, weeding operations must be scheduled to avoid competition overtaking and suppressing tree growth. Other operations, such as pruning, need to be scheduled depending on the pruning height requirements, species, and quality specifications.

1.9.4 Operational plan

Stand-level planning of how labour and equipment will be used to execute the tasks. This daily plan is derived from the APO and is activity-specific. The specific area to be worked on is planned in detail by:

- Conducting a risk assessment
- Using a map to indicate the operational components of the specific operation
- Limitations of equipment indicated on the map (slope, ground roughness, or slope)
- Labour requirements
- Machine requirements and application
- Resource requirements and application procedures (e.g. chemicals, fertiliser, stock absorb, and so on)
- Any sensitive areas are indicated.

1.10 Summary

This chapter provides an overview of global plantation forestry, with a focus on South Africa and Brazil. Furthermore, it highlights the importance of sustainable forest management in meeting the increasing global demand for wood. It examines forest composition and plantation species in South Africa and Brazil, covering land use, species distribution, and production volumes.

It discusses silviculture practices, particularly in clear-fell even-aged systems, and the role of technology in silviculture, while addressing challenges in adopting machinery and modern technology. It also discusses the importance of silviculture, its evolution driven by various factors, including improvements in health and safety, increased operational productivity, reduced operational costs, enhanced stand productivity, and social and environmental aspects. This chapter also discusses strategic, tactical, and operational planning in silviculture.

Review questions

1 What role do plantation forests play in society?
2 Which silvicultural system is used in South Africa and Brazil?
3 How has silviculture technology progressed over the years?
4 Describe the levels of silviculture technology classification
5 Describe the levels of silviculture operations planning

References

Achim, A., Moreau, G., Coops, N. C., Axelson, J. N., Barrette, J., Bédard, S., Byrne, K. E., Caspersen, J., Dick, A. R. & D' Orangeville, L. 2022. The changing culture of silviculture. *Forestry,* 95, 143–152.

Branca, G., Piredda, I., Scotti, R., Chessa, L., Murgia, I., Ganga, A., Campus, S. F., Lovreglio, R., Guastini, E. & Schwarz, M. 2020. Forest protection unifies, silviculture divides: A sociological analysis of local stakeholders' voices after coppicing in the Marganai forest (Sardinia, Italy). *Forests,* 11, 708.

Brasil Ministério da Agricultura e Pecuária. 2024. *Plano Nacional de Desenvolvimento de Florestas Plantadas – PNDF. Atualização: março de 2024* [Online]. Brasilia: Ministério da Agricultura e Pecuária, 24 p. Available: https://www.gov.br/agricultura/pt-br/assuntos/sustentabilidade/floresta-plantada/publicacoes/plano-nacional-de-desenvolvimento-de-florestas-plantadas/ [Accessed 8 August 2025 2024].

Brazilian Tree Industry. 2023. 2023 IBÁ Annual Report. *In:* Tech, E. (ed.). IBÁ. Available: https://iba.org/wp-content/uploads/2025/05/relatorio-anual-iba2023-r.pdf [Accessed 7 December 2025].

Brazilian Tree Industry. 2024. 2024 IBÁ Annual Report. *In:* Tech, E. (ed.). IBÁ. Available: https://iba.org/wp-content/uploads/2025/05/relatorio2024.pdf [Accessed 7 December 2025].

Chapman, G. W. & Allan, T. 1978. *Establishment Techniques for Forest Plantations*. Rome: FAO.

Choudhry, H. & O' Kelly, G. 2018. Precision forestry: A revolution in the woods. *Advanced Technologies Could Improve Forest Management Significantly. What Areas Are Most Promising, and How Can Forestry Companies Start Their Digital Transformation?* Singapore: McKinsey&Company, 8–10. Available: https://www.mckinsey.com/industries/packaging-and-paper/our-insights/precision-forestry-a-revolution-in-the-woods#/ [Accessed 7 December 2025].

da Silva, R. B. G. 2023. Manejo de irrigação em viveiros florestais: uso racional da água e análise da qualidadeDoss, C. R. 2006. Analyzing technology adoption using microstudies: Limitations, challenges, and opportunities for improvement. *Agricultural Economics,* 34, 207–219.

Erasmus, D. 1994. National terrain classification system for forestry. *ICFR Bulletin Series No. 11/1994.* Pietermaritzburg: Institute for Commercial Forestry Research.

Estatística, I. B. D. G. E. 2025. *PNAD Contínua: Pesquisa Nacional por Amostra de Domicílios Contínua: principais destaques.* Rio de Janeiro: IBGE.

Evans, J. & Turnbull, J. W. 2004. *Plantation Forestry in the Tropics: The Role, Silviculture, and Use of Planted Forests for Industrial, Social, Environmental, and Agroforestry Purposes.* Oxford and New York, NY: Oxford University Press.

FAO. 2020a. *Global Forest Resources Assessment 2020: Main report.* Rome: FAO.

FAO. 2020b. Global Forests Resources Assessment 2020. *Terms and Definition FRA 2020.* Rome: FAO.

FAO. 2022. *Global Forest Sector Outlook 2050: Assessing Future demand and Sources of Timber for a Sustainable Economy – Background Paper for The State of the World's Forests 2022.* Rome: FAO.

Forestry South Africa. 2023. *22nd Annual Report for Year Ended 31st December 2023.* Pietermaritzburg: Forestry South Africa.

Freer-Smith, P. H., Muys, B., Bozzano, M., Drössler, L., Farrelly, N., Jactel, H., Korhonen, J., Minotta, G., Nijnik, M. & Orazio, C. 2019. *Plantation Forests in Europe: Challenges and Opportunities.* Joensuu: European Forest Institute Joensuu.

FSA Environmental Management Committee (ed.) 2019. *Environmental Guidelines for Commercial Forestry Plantations in South Africa.* Birnam Park: Forestry South Africa.

Gamborg, C. & Larsen, J. B. 2003. 'Back to nature'—a sustainable future for forestry? *Forest Ecology and Management,* 179, 559–571.

Godsmark, R. & Oberholzer, F. 2020. The South African forestry and forest products industry 2019. *Annual Stats of SA Forestry Industry.* Pietermaritzburg: Forestry South Africa.

Guerra, S. P. S., Oguri, G., Schumacher, M. V., Kulmann, M. S. D. S. & Faria, J. I. 2023. *Survey of the Level of Mechanization in Forestry: 2022/2023 Edition.* Piracicaba: IPEF.

Kelty, M. J. 2006. The role of species mixtures in plantation forestry. *Forest Ecology and Management,* 233, 195–204.

Kotovic Filho, J. 2016. Mecanização das atividades de silvicultura em empresas florestais brasileiras. 57 f. Curitiba: *Monografia (Graduação em Engenharia Florestal)-Universidade Federal do Paraná.*

Kratzer, J., Meissner, D. & Roud, V. 2017. Open innovation and company culture: Internal openness makes the difference. *Technological Forecasting and Social Change,* 119, 128–138.

Li, G., Hou, Y. & Wu, A. 2017. Fourth industrial revolution: Technological drivers, impacts and coping methods. *Chinese Geographical Science,* 27, 626–637.

Liu, C. L. C., Kuchma, O. & Krutovsky, K. V. 2018. Mixed-species versus monocultures in plantation forestry: Development, benefits, ecosystem services and perspectives for the future. *Global Ecology and Conservation,* 15, e00419.

Matthews, J. D. 1991. *Silvicultural Systems.* Oxford: Clarendon Press.

Nyland, R. D. 2016. *Silviculture: Concepts and Applications.* Long Grove, IL: Waveland Press.

Oberholzer, F. 2021. The South African forestry and forest products industry 2019. *Annual Stats of SA Forestry Industry*. Pietermaritzburg: Forestry South Africa, 1–160.

Oberholzer, F. 2023. The South African forestry and forest products industry 2019. *Annual Stats of SA Forestry Industry*. Pietermaritzburg: Forestry South Africa, 1–160.

OMNRF. 2015. *Forest Management Guide to Silviculture in the Great Lakes-St. Lawrence and Boreal Forests of Ontario*. Toronto: Queens Printer for Ontario.

Pott, R. M. 1997. Plantation forestry in South Africa and its impact on biodiversity and water. *Southern African Forestry Journal,* 180, 45–48.

Puettmann, K. J., Wilson, S. M., Baker, S. C., Donoso, P. J., Drössler, L., Amente, G., Harvey, B. D., Knoke, T., Lu, Y. & Nocentini, S. 2015. Silvicultural alternatives to conventional even-aged forest management-what limits global adoption? *Forest Ecosystems,* 2, 8.

Rubilar, R. A., Allen, H. L., Fox, T. R., Cook, R. L., Albaugh, T. J. & Campoe, O. C. 2018. Advances in silviculture of intensively managed plantations. *Current Forestry Reports*, 4, 23–34.

Ramantswana, M., Brink, M., Little, K., Spinelli, R. & Chirwa, P. 2021. Perspectives and drivers of modernisation of silviculture re-establishment in South Africa. *Southern Forests: A Journal of Forest Science,* 83, 79–87.

REN. 2022. *Renewables 2022 Global Status Report*. Paris: REN21 Secretariat.

Roberts, R., Flin, R., Millar, D. & Corradi, L. 2021. Psychological factors influencing technology adoption: A case study from the oil and gas industry. *Technovation,* 102, 102219.

Sands, R. 2013. Forestry in a Global Context. Wallingford: CABI.

Savill, P., Evans, J., Auclair, D. & Falck, J. 1997. Plantation Silviculture in Europe. Oxford: Oxford University Press.

Scotcher, J. 2006. Forest certification in South Africa. *South African Forestry Journal,* 206, 1–3.

Seenivasan, R. & Parthiban, K. 2017. Precision silvicultural technology for industrial plantations. *Forestry Technologies-A Complete Value Chain Approach*, 194–216.

Shepherd, K. 1986. *Plantation Silviculture*. Dordrecht: Martinus Nijhoff Publishers.

Smith, D. 1986. *The Practice of Silviculture*. 8th ed. New York: John Wiley.

Soler, R. R. 2020. *Desempenho operacional e econômico do ensaio de uma plantadora tripla automatizada de mudas florestais.* PhD thesis, Universidade Estadual Paulista (UNESP).

Statistics South Africa 2019. *Quarterly Labour Force Survey Quarter 1: 2019*. 14 May 2019 ed. Pretoria: Stats SA.

Stephens, S. S. & Wagner, M. R. 2007. Forest plantations and biodiversity: A fresh perspective. *Journal of Forestry,* 105, 307–313.

Strategy Execution Advisers 2020. *Masterplan for the Commercial Forestry Sector in South Africa: 2020–2025*. Pretoria: DEFF.

Szmyt, J. & Dering, M. 2024. Adaptive silviculture and climate change—A forced marriage of the 21st century? *Sustainability,* 16, 2703.

United Nations. 2023. *Our growing population* [Online]. United Nations. Available: https://www.un.org/en/global-issues/population#:~:text=The%20global%20human%20population%20reached,and%202%20billion%20since%201998. [Accessed 16 February 2023 2023].

Uusitalo, J. & Pearson, M. 2010. *Introduction to Forest Operations and Technology*. Hämeenlinna: JVP Forest Systems Oy Hämeenlinna.
West, P. W. 2014. *Growing Plantation Forests*. Cham: Springer.
Zhang, T., Niinemets, Ü., Sheffield, J. & Lichstein, J. W. 2018. Shifts in tree functional composition amplify the response of forest biomass to climate. *Nature,* 556, 99–102.
Zwolinski, J. & Bayley, A. 2001. Research on planting stock and forest regeneration in South Africa. *New Forests,* 22, 59–74.

2 Nursery technologies

Muedanyi M. Ramantswana,
Saulo P. S. Guerra, and
José Mauro S. da Silva

2.1 Introduction

The successful establishment of forest plantations depends on high-quality plant material. This chapter focuses on various nursery technologies, exploring the methods and advancements essential for the propagation of plants used in the regeneration of plantation forests in South Africa and Brazil. In this chapter, the reader will learn about various nursery phases, from rooting and growing to hardening off and dispatch, emphasising the importance of watering, fertigation, and weeding practices. Furthermore, this chapter discusses plant quality assessment parameters, with a focus on seedlings suitable for mechanised planting, and examines existing and upcoming precision technologies.

2.2 Nursery practices in South Africa and Brazil

Nursery material is derived from tree breeding selection processes, enabling individual trees with the required traits to be propagated. Genera such as *Eucalyptus*, *Pinus*, and *Acacia* are propagated using different methods, including seeds and cuttings. Seeds derived from relatively controlled pollinated trees. Before the seeds are used in the nursery, they are harvested from seed orchards, and their seed moisture is tested. Germination tests are then conducted, after which the seeds are packaged and stored. When required, the seed undergoes pretreatment before direct sowing for seedling production at the nursery (Shepherd, 1986b). The comprehensive genetic theory and practice are beyond the scope of this book. However, the material (seed or cuttings) used for nursery propagation must be derived from good trees with vigorous growth, healthy trees, straight stems, thin and wide branches, and good wood quality (van Wyk and Verryn, 2000).

In Southern Africa and Brazil, nursery technologies vary from manually oriented facilities producing small quantities to more advanced, highly technical and automated nurseries which produce large quantities to meet the forestry industry's needs. Most nursery environments are semi-controlled (e.g. shade houses) or fully controlled (e.g. greenhouses).

DOI: 10.4324/9781003391227-2

Fully controlled environments are preferable because plants can be grown faster and all year round; however, they are more costly to construct (Dumroese et al., 2009). The two primary methods of raising seedlings in a nursery setup are bare-rooted and containerised (Evans and Turnbull, 2004). Bare-root seedlings are grown in open fields characterised by moist soil conditions, and the seedlings are planted free from the soil. In contrast, containerised systems are grown in containers with artificial growing media. The root plug comprises growing media retained from the nursery until the seedling is planted (Grossnickle and El-Kassaby, 2016). This book will focus on technologies used in planting/sowing, growing, tending, and dispatching plants in a containerised nursery, the predominant way plants are produced in South Africa and Brazil (Zwolinski and Bayley, 2001). In plantation forestry applications, containerised plants are more advantageous than bare-root systems because they are (Grossnickle and El-Kassaby, 2016, Idris et al., 2004):

- Cost-effective and efficient, depending on location and resources
- Shorten the production cycle
- Enhanced plant quality
- Extended planting period
- Improve seedling performance (e.g. survival), especially in harsh conditions
- Root system protection during storage, transportation, and transplanting

2.3 Growing media

Various sources of growing media are available to grow container seedlings in the nursery. Growing media can be referred to as a substrate or a soilless medium. Good media needs to provide physical support to the plant, adequate aeration, good water holding capacity and supply mineral nutrients (Landis et al., 1990). The growing medium may comprise soil, *Sphagnum* peat moss, coconut coir, compost, softwood pine bark, and wood fibre (Barrett et al., 2016). Furthermore, perlite and vermiculite can be added to the potting mix to improve aeration and drainage (Shepherd, 1986b, Mohanan and Sharma, 2005). Over the past few decades, coir, pine bark, and wood fibre have become preferred renewable alternatives to peat due to performance, economic and environmental drivers (Barrett et al., 2016, Riikonen and Luoranen, 2018). In South Africa, the most common growing medium used to produce hardwood and softwood plants was composted pine bark (Mitshell, 2012). However, there has been a change in combining pine bark with coco-peat and coir fibre material. Producing this medium involves ageing the bark through hammer milling and allowing for decomposition until a pH of around 5.5. When ready for use in trays, it should have a maximum particle size of 12 mm, be black, and be able to retain water after irrigation (Mitshell, 2012). In Brazil, large-scale forestry nurseries adhere to similar practices.

2.4 Containerised systems

Nursery plants are raised in a growing medium and placed in a container or tray. In South Africa, all planting stock is currently produced from containerised nursery systems (Zwolinski and Bayley, 2001). The containers can have loose or fixed cavities (inserts) from moulded plastic or polystyrene material. The container should support the plant and facilitate good rooting development (Shepherd, 1986b) because the plants are raised separately from other plants. In Brazil, rigid plastic tubes are widely used due to their durability and ease of handling, favouring the formation of a more orderly root system. However, biodegradable alternatives, such as pots made of recycled paper or plant fibres, have gained prominence for reducing environmental impact and eliminating the unpotting step (Cury et al., 2024, Silva et al., 2025).

During the establishment phase, containers are used as a carrier of seedlings from the nursery to the stand where they are planted. In semi-mechanised and mechanised planting operations, the type of containers used can influence both the productivity of the planting operations and the survival of the plants. Containers are available in various shapes and sizes, manufactured from different materials, and each design has its benefits and disadvantages (Schuermans, 2011). In South Africa, large grower companies and private nurseries use different types of containers, with the most common ones being the Unigro 128, Unigro 98 tray and polystyrene moulded trays (Williams, 2000) (Table 2.1). In Brazil, the most common trays are the last four, with the last one being used for biodegradable products. Containers should have good drainage, allow root pruning (air or manual), root guiding, and vertical ridges to direct the roots downwards (Evans and Turnbull, 2004, Almeida, 2016).

2.5 Nursery production process flow

This section explains the process and technologies used for the production of plants in a forestry nursery.

Table 2.1 Types of containers used in forestry

Container type	*Number of cavities*	*Tray size (mm)*	*Cavity volume (cm^3)*	*Cavity length (mm)*	*Density (cells per m^2)*
Unigro 128	128	676 × 350 × 100	65	100	540
Unigro 98	98	676 × 350 × 100	90	100	414
Ellepot-125	128	676 × 350 × 100	60	90	540
Inert frame-128	128	676 × 350 × 100	85	100	540
Flat	176	620 × 420 × 160	55	35	704
Box	187	620 × 420 × 160	55	35	748
Box	54	625 × 425 × 165	280	65	216
Paper pot	126	580 × 376 × 78		36	578
Paper pot	176	585 × 377 × 83	56	33	704

The first four are in South Africa, and the last one is in Brazil.

2.5.1 Plant propagation using seed and cuttings

The nursery production process starts with seed or cuttings (shoots) being sown or set directly into the growing medium. Most common large-scale production nurseries use the sectored system where plants are moved to different locations depending on the development phases, such as germination/rooting, growing and hardening (Stape et al., 2001). In South Africa, pine seedlings are propagated from open-pollinated seed derived from genetic tree improvement programmes established and managed by large corporate companies over several decades of research (Mitshell, 2012). Vegetative propagation can be conducted using grafting, rooted cuttings, air layering and tissue culture (Naidoo, 2012). This book predominantly focuses on the use of seed and rooted cuttings raised in containerised systems. The cuttings are harvested manually from established hedge or clone banks using scissors to cut the growing tips and secateurs for hedge shaping. There is a trend wherein outdoor field hedges are shifting to indoor hydroponics (fertigated sand beds) mini hedges, which have enhanced shoot growth (Evans and Turnbull, 2004) (Figure 2.1). Furthermore, this indoor system ensures that the nutrition, temperature, and light are well controlled, which produces high rooting success (de Assis et al., 2004).

2.5.2 Production house

2.5.2.1 *Container filling*

Containers are traditionally filled manually with the help of a funnel (Figure 2.2). The coco-peat coir, typically supplied in blocks, is broken into finer material by a coir milling machine and moisturised (adding water). Once the coir is ready, it can be manually or mechanically loaded

Figure 2.1 Hedge plants harvested for cuttings

Figure 2.2 Manual container filling

onto a conveyor belt, and an additive like perlite or vermiculite is supplied separately to the soil mixer. The proportions of growing medium and additives will differ depending on the species to be planted. Once the growing medium and additives are in the soil mixer, controlled-release fertiliser (e.g. Osmocote) can be added and mixed. Using a conveyor system, the soil mixer is used as a feed hopper to supply the enhanced growing medium to a tray-filling machine. In advanced nurseries, the container filling process is performed by a tray filler, which can be used as a stand-alone or integrated with a working line (Urbinati, 2024). Depending on which tray-filling system is integrated, the growing medium can be supplied to a loose-filled container machine or a paper-pot-producing machine. Both countries are shifting from using loose-filled containers to paper pot systems.

When using the paper pot machine, the propagation paper is made of wood fibres and is degradable (Ellepot, 2024). The paper rolls are loaded on the machine and automatically unwind as the machine operates. The side edges of the paper are glued together as the growing medium is supplied, filling the roll, which is then consistently cut into required sizes before being deposited into the container. According to Naidu et al. (2022), operationally, paper pots are advantageous compared to conventionally loose-filled because:

- Paper is biodegradable
- They allow roots to penetrate the paper and get air-pruned
- Better survival of the plant
- Decreased utilisation of plastic inserts used with conventional trays
- Improved tray logistics, i.e. transporting, cleaning, and packing of trays
- Improved structure of root plug and substrate

2.5.2.2 Setting of cuttings

The vegetative propagation of eucalyptus species is used extensively. Mini-cuttings ranging between 4 and 10 cm long are harvested from the hedge plants. The recommended cutting length for eucalyptus hybrid clones in South Africa is 8–10 cm for optimal rooting (Naidu and Jones, 2009). In Brazil, authors cite mini-cuttings with around 6–10 cm in length (Borges et al., 2011).

Every eucalypt cutting planted into the container growing media should have a pair of leaves with about 50% of the leaf blade cut off (Maile and Nieuwenhuis, 1996, de Assis et al., 2004, Brondani et al., 2010). Softwood cuttings comprising lateral branches (6–8 cm from apical bud) are harvested from actively growing hedges or stock plants. After that, the cuttings are inserted (not more than 2 cm) into preprepared dibbled holes filled with moist growing media in the container (Tousignant et al., 2007, Mitshell, 2012). Setting cuttings is still a manual process involving labour responsible for harvesting and setting the cuttings directly into the containers (Figure 2.3). After harvesting

Figure 2.3 Manual setting of cuttings

the cuttings from the hedge plants, they are temporarily kept in a cooler box and transported to the setting shed to avoid wilting and desiccation before being planted into the containers. New technology that automatically identifies cuttings spread on a conveyor is being tested. The robot picks one cutting and consistently plants it at the required depth to improve the uniformity of the plants. The application of self-learning software enables the identification and distinction of different species of plants that have to be planted in the tray containing a particular substrate, such as loose soil, peat or paper plugs (ISO, 2023).

2.5.2.3 Sowing seed

Seeds can be sown by hand. The number of seeds sown per cavity depends on a particular species and the expected seed germination percentage. The seed is planted at a depth of two to three times the width of the seed and immediately covered by a substrate (e.g. composted bark, perlite, vermiculite) (Landis et al., 2008).

Machines such as the semi-automatic seeder can be used to improve the manual, labour-intensive seeding work. Before sowing the seeds, compactors push the growing media down by 1 cm, creating a space for seeds to be deposited into the cavities (Liegel, 1987).

Modular seeding machines with automatic drum and row seeding lines are suitable for use in modern nursery operations. The drum seeding machine utilises an advancement belt that carries a container filled with growing medium. The container is conveyed to the dibbler, which has conical pins which create precise (same depth and position) indentations in every cell of the container. After that, the container proceeds to the sowing drum, which is attached to a vacuum pump and compressor. The rotating drum is under vacuum and has a row of holes so that the seeds can be sucked onto the drum; as the drum continues rotating, the bottom centre of the row of holes with seeds attached changes from vacuum pressure to releasing the seed to the container (tray) (Urbinati Nursery Technology, 2025). The moving container is synchronised to the speed of the rotating drum to ensure that the seed is deposited precisely where it needs to be in the tray. The container with seeds moves to the covering unit, which releases vermiculite to cover the seed (Urbinati Nursery Technology, 2025).

2.5.2.4 Watering

Sown seeds and set cuttings require adequate moisture for their survival. According to Dumroese et al. (2009), the irrigation requirements can differ depending on the seedling's growth phase (establishment, rapid growth, and hardening). Watering during the establishment phase must be applied to replenish the moisture in the growing medium's top layer; this can be achieved by misting or regular water application using a fine mist nozzle and low flow rate. During the rapid growth phase, more frequent watering is required as the plant's root system expands and requires growing resources. Higher-quality seedlings are those that survive and grow faster after planting. This quality, expressed through morphological and physiological attributes, can be achieved through nursery management, including water management. Using higher-quality seedlings reduces the need for replanting and the costs of cultural practices in the initial months after planting (da Silva, 2023). Furthermore, the application of liquid fertilisers supplements growth; however, it may lead to increased salts; hence, regularly flushing the irrigation system with clean water is essential.

During the hardening phase, water stress can be induced by limiting watering to prepare the plant for the natural environment (Dumroese et al., 2009). Leaf water potential, stomatal conductance, relative water content in the leaves, daily transpiration, leaching fraction, and total dry mass were evaluated. In tree species with positive mean leaf angles, lower irrigation depths are sufficient to enhance leaf water potential, stomatal conductance, relative water content, and transpiration efficiency. In contrast, for seedlings with negative mean leaf angles, higher irrigation depths are required to maintain adequate physiological responses and avoid reductions in total dry mass production (Delgado et al., 2022).

Manual watering with a hose produces uneven watering. Nurseries commonly use overhead irrigation systems, which can be fixed sprinklers or moving booms. For precise and automated watering, a sprinkler system can adequately provide plant moisture as and when needed (Evans and Turnbull, 2004). According to Kruger (2012), boom irrigation systems are the most effective because of their improved precision when spraying. Irrigation systems can be automated using time clocks and container weight monitoring technologies. Most nurseries in South Africa employ a timer-based irrigation system, which irrigates the growing media close to the saturation point without considering the plant's specific water needs (Donald, 1986). Automated irrigation systems can be successfully used for scheduling irrigation of eucalypt seedlings based on measurements of the media water content provided by EC-5 soil water content sensors (Kaptein et al., 2019). Other research has found that when the weight of a sample group of containers reaches a specific level due to transpiration loss, the watering device can be activated to supply the required water (Shepherd, 1986a). Applying technology, such as radio telemetric measurement systems, improves water use efficiency and automates the irrigation process at every stage of seedling growth (Durło et al., 2018).

2.5.2.5 Fertilisation

The fertilisation of plants at the nursery must be considered after considering nutrients available from the growing medium, the water's chemistry and the plants' nutritional needs, depending on the growth stage (Kruger, 2012). Slow-release fertilisers can be incorporated into the growing medium to fill the trays. Overall, plants require a regular and good balance of nitrogen, phosphorous, and potassium (NPK) and micronutrient mixture application during the early and growing phase to support growth; however, nitrogen application is mostly restricted during the hardening phase to restrict rapid plant growth (Stape et al., 2001).

The most common technology used to apply fertiliser is through the fertigation method. This method implies that the irrigation system supplies soluble fertilisers to the plants. Advanced fertigation systems use pumps to release accurate quantities (ratios) of fertiliser into the irrigation line as needed. Fertigation can occur using the automated hydraulic boom sprayer, which can apply the solution uniformly across the seedlings (Dumroese et al., 2009). The advantage of using fertigation systems is that precise concentrations of nutrients can be applied, the nutrient solution can be changed depending on plant needs, and there is a lower risk of over-fertilisation and salt damage to plants (Landis and Nisley, 1990).

2.5.3 Germinating or rooting phase

To facilitate successful rooting, the containers with the seeds or cuttings are moved and placed in the rooting tunnel, which must have a controlled temperature (winter and summer), good aeration and adequate irrigation (Figure 2.4). Temperature and humidity control are essential factors to monitor during this phase. For advanced nurseries, accurate real-time management of these variables is achieved by heating the floors during colder periods and utilising either a greenhouse ventilation system (e.g. Butterfly roof ventilation) or an evaporative cooling system with a pad and fan. Good drainage can be achieved through concrete floors; however, gravel floors are often more cost-effective and commonly used. Seedlings grown in containers must develop a root system that fills and holds the growing medium, consolidated

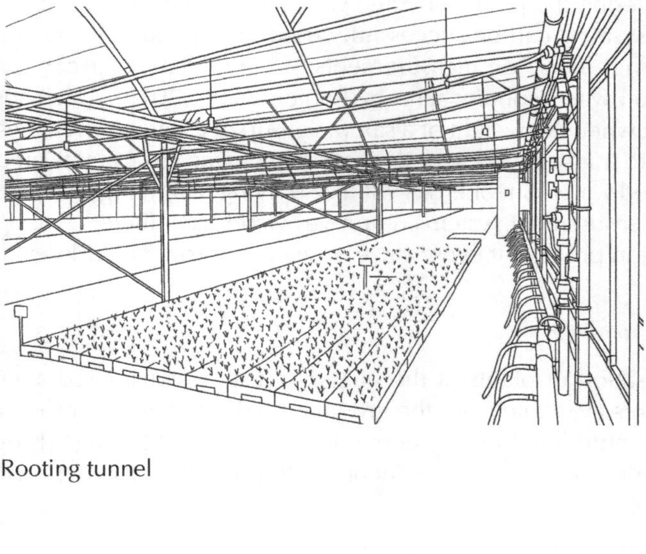

Figure 2.4 Rooting tunnel

to ensure the plug does not quickly disintegrate (Grossnickle and Ivetić, 2022). Within the rooting greenhouse area, if bottom heating is installed, the rooting period for cuttings can be shortened by up to 50% (Williams, 2000); heating systems are mainly applied during the colder winter period. Gas boilers can supply the needed heat in the rooting camp. A ground-floor temperature of around 22°C is maintained to encourage rooting. For irrigation, mist sprayers or spray nozzles installed in the rooting tunnel structure are set to distribute fine water droplets consistently during the rooting phase (Mitshell, 2012). For example, misting can be activated in some hardwood rooting camps every seven minutes for 20–30 seconds, depending on the season. Depending on the species and season, hardwood cuttings can be kept in the rooting camp for a period ranging between four and seven weeks. Visual observations are conducted regularly to monitor the development of rooting. Once the cuttings have been rooted, the management systems tend to grow the plants in a similar way to that of seedlings (Kruger, 2012).

2.5.4 Growth phase

The growing phase is a time when the plants experience rapid growth and development. The containers are placed on elevated rails (Figure 2.5). Higher nitrogen levels are supplied through a fertigation system to encourage growth. During this phase, seedlings can be graded, sorted, and topped. Grading refers to dividing the plants into different classes, while sorting is the process of selecting acceptable plants and removing the unacceptable plants (Poon, 1996). In conventional nurseries, seedlings are graded and sorted manually (Rigney, 1986) through observation based on specific criteria requirements. However, there is evidence that machine vision can be accurately assessed based on specific criteria such as root collar diameter, height and root area development (McGuinness et al., 2021, Poon, 1996).

2.5.4.1 Chemical and fungicide application

Weeds in a nursery can compete with the seedlings for water, nutrients, and light; hence, monitoring and controlling weeds is essential throughout all the nursery phases (Vasic et al., 2012). Weeds are not a significant problem in nurseries because of the containerised system and relatively clean growing medium (Donald, 1986). Manually weeding the containers can be performed by hand. Chemical weeding is more efficient, even though its application is limited to specific, selective herbicides (Evans and Turnbull, 2004). There is increasing environmental pressure to significantly reduce the application of chemicals and fungicides in nursery environments (Okorski et al., 2015). Among the pathogens responsible for seedling damping-off, the fungus *Sclerotium rolfsii* Sacc. is one of the most notable pathogens associated with seedling damping-off. This is primarily due to its vast host range – affecting nearly 500 plant species with high survival capacity in the soil, and its wide

Figure 2.5 Growing tunnel

geographic distribution, mainly in tropical and subtropical regions under conditions of high humidity and elevated temperatures (Mafia et al., 2007). According to Evans and Turnbull (2004), hand weeding will probably continue to be the primary weed control method in tropical nurseries because it is safe, easy, and appropriate for all species. With the advent of enhanced artificial intelligence, deep learning (Wang et al., 2019, Harders et al., 2023), and robotics, automated weed detection and removal may be integrated into future nurseries (Garrido et al., 2014).

According to Okorski et al. (2015), modern nurseries face challenges of increasing varieties of diseases driven by high moisture and nutrient supply conditions. Some common diseases in nurseries include *Fusarium* spp. (seed rot, damping off, root diseases, stem blight), *Phytophthora spp.* (root disease), and *Pythium* spp. (root disease), *Botrytis cinerea* (foliar blight, storage mould),

Cylindrocarpon destructans (root decay) (Rajković et al., 2007). The application of fungicides is currently an effective technique to limit the spread of pathogens. However, integrating pest management (IPM) strategies that seek to reduce pesticide application and provide substitutes to deal with diseases effectively must be prioritised (Poteri et al., 2005). The application of cultural (e.g. sanitisation) and biological control (e.g. biofungicides and biocontrol agents) measures is a vital alternative to consider where conditions permit (Rajković et al., 2007).

2.5.5 Hardening off

The hardening off phase is the final stage at the nursery before the seedlings are dispatched to the forest. At this stage, plants are typically placed on raised benches, allowing for adequate airflow and root pruning while having a weed- and algae-free area covered by gravel or concrete surfaces (Nelson, 2012). Hardening plants at the nursery aims to increase their durability and resistance to stresses imposed by the natural environment. Well-hardened plants are essential for semi-mechanised and mechanised planting operations because they are exposed to various shocks from the nursery until they are finally planted in the field. The hardening of seedlings may include reducing water, certain fertilisers and shade (to influence light and temperature) (Land Jr and Cunningham, 1994, Dumroese et al., 2009). However, a significant correlation exists between seedling survival and their stem diameter at the end of the nursery phase, highlighting the importance of this trait for successful field establishment, regardless of the substrate used (Stuepp et al., 2020). Seedlings with a higher-quality root system have a greater capacity to overcome the initial phase of planting at the planting site (da Silva, 2023). The morphological plant features, such as stem diameter and root system, can affect the ability of a plant to survive after being planted in the field. If the nursery plants are growing too tall for the root system, the shoots can be pruned by cutting the succulent tops with scissors to ensure the plants are uniform. Uniformity (e.g. height, root to shoot ratio) of plants to be out-planted is necessary in semi-mechanised and mechanised operations because much variability (plant form and height) may cause blockages in planting tubes and damage to the plant itself.

2.5.6 Plant dispatch and logistics

Most plants are dispatched and delivered to the plantation in the containers used to grow them. Before being dispatched, they are sorted, graded, and assessed to ensure they meet plant specifications required by the client. Grading seedlings must be performed at the nursery, where trained staff can consistently select the seedlings that meet specifications before dispatching them to the customer (Figure 2.6). Visually inspecting and culling poor seedlings, such as those that are damaged, with poor colour, and diseased seedlings, needs to be conducted comprehensively (Evans and Turnbull, 2004).

Figure 2.6 Manual grading and sorting of seedlings before dispatch

Advanced technological systems that use artificial intelligence (e.g. computer vision), which enable the rapid assessment of plant quality, can be used for plant quality assessments. For example, using a mobile device camera and programming, one can automate plant height measurements within a nursery tray. This data can provide information to develop growth curves and assist with automated measurement of plant quality. Other sensing technologies, such as multispectral and thermal imaging, can monitor issues such as plant health and water stress (Naidu et al., 2022).

To transport planted seedling trays within the nursery and the dispatch area, tractors with towing self-steering trailer systems can efficiently move the trays around confined areas (Nelson, 2012). Samples of trays are checked before dispatch to determine the number of plantable seedlings per tray. When delivering seedlings, the area must be enclosed (e.g. with a tarp) to protect the seedlings from wind and sun damage (Liegel, 1987). The trays must be stable, secured, and easily slid in and out of the rail structures on the truck. If loading or offloading is done manually, a stable platform with side railings must be used to ensure worker safety when working at heights.

It is essential to note that trays dispatched with seedlings are unavailable for the nursery to use until they are returned after the seedlings have been planted. Trucks that deliver full seedling trays often return to the nursery with a load of empty trays. An up-to-date reconciliation of trays taken out and returned is essential, as tray inventories can account for large amounts of money in nurseries. The trays are usually loaded in an enclosed, rigid truck (sometimes with a drawbar trailer). The truck bed can have several tiers with adequate spacing to avoid seedling damage.

Other methods are used globally to store and deliver plants to the forest site, such as jelly rolling, bagging, and boxing. These methods facilitate a more straightforward mechanised process through automatic bagging machines (Landis et al., 2010). A rack-type system is preferred for transporting seedlings as it avoids plant damage and facilitates more effortless mobility (Jooste, 2022).

Plantation forest areas commonly use holding nurseries as temporary plant storage, before being transported to the infield (Hechter et al., 2022). The holding nursery should have adequate irrigation, enable plant hardening, and not create conditions that allow pathogens to attack the plants. Ideally, plants should not be kept at the holding nursery for more than a week before being planted in the infield (Jooste, 2022).

2.6 Automation technologies at the nursery

Nurseries in South Africa and Brazil remain heavily reliant on labour for plant propagation. However, various technologies are being introduced predominantly by large corporate and private nurseries to improve efficiencies and reduce dependence on labour. Most advanced technologies described in this section have not yet been widely adopted; however, they are being tested and refined and may be introduced on a large scale in the future due to their reliability, accuracy, and cost-effectiveness. Sensing and automation technologies in nurseries are becoming more prevalent in operations such as smart irrigation, plant stress detection, variable spray applications and plant biometric measurements (Mahmud et al., 2023). Automation of nurseries can reduce dependence on manual labour and decrease chemical, pesticide and irrigation costs through precise and efficient application of resources (Rihn et al., 2023). Using sensors to detect specific parameters and communicate with various controllers is vital in improving efficiencies in forest nurseries. For example, soil moisture in a containerised nursery can be monitored using probes in the growing medium, cameras can be used to assess plant growth characteristics, and irrigation can be activated according to each plant's needs.

2.6.1 Precision irrigation, fertigation, and climate management

Nurseries utilise automated irrigation, fertigation, and greenhouse climate control. When it comes to irrigation, most nurseries tend to over-irrigate; however, the use of real-time moisture assessments can facilitate automated and accurate irrigation, ensuring water is provided when needed and in the correct quantity as required by the plants. The two most common technologies used to automate irrigation in forest nurseries are sensor readings of the soil/substrate water status, plant water status or evapotranspiration rates derived from weather conditions and species-specific factors (Van Iersel et al., 2013). Technologies such as moisture probes (Durło et al., 2018b) and gravimetrical methods (e.g. scale) (Girard and Gagnon, 2016) can be used to monitor the water content in several trays and regularly communicate the data to a computer system with software that determines the moisture and triggers the irrigation system to supply adequate watering until the specified moisture is achieved.

Climate control automation in containerised forest nurseries, which grow plants in greenhouses, involves the automatic management of air temperature, air circulation and exchange, relative humidity, and light (Shamshiri and Ismail, 2013). Conventional greenhouses have sensors inside and outside that monitor critical environmental variables that directly influence seedlings' growth and growth conditions. For instance, to control temperature, the ventilation systems of the greenhouse need to be managed. This can be controlled automatically using different evapotranspiration cooling methods such as pad and fan systems, unit coolers and misting systems (Shamshiri and Ismail, 2013). The required conditions are set or preprogrammed to the optimal conditions suitable for plant growth at different growth stages. A control system can comprise an industrial computer with its electronic system and an interface that visualises the irrigation, with pumps, filters, fertiliser tanks, and piping layout. This system can be programmed to change, monitor and retrieve irrigation, fertilising, and climatic settings at any given time. Furthermore, the system can be controlled remotely from an electronic device such as a smartphone, personal computer, or computer (Hermisan, 2024).

2.6.2 Plant quality assessments

Forest nurseries must meet specific plant quality requirements that can be assessed and quantified. Plant quality assessments commonly assessed are morphological and physiological attributes. Plant quality assessment is conducted manually through visual observation and measurements (using rulers). Each species has plant quality specifications checked against a specific score sheet before being dispatched (Jooste, 2022, Ford, 2013). The specifications include the age of plants, plant size, root plug colonisation, root spiralling/coiling, plant health (leaf colour, mycorrhiza and disease), and morphological attributes, including seedling height, diameter and root development (Ford,

2013, Faustino et al., 2022, Lopes et al., 2016). The physiological characteristics assessed include nutrient and water status (Grossnickle and MacDonald, 2018). The Dickson Quality Index (DQI) is widely used in Brazilian research as an integrating parameter for evaluating the quality of forest seedlings. It simultaneously considers the total biomass, the distribution between aerial and root parts, and the robustness of the seedling. DQI is one of the most reliable indicators for predicting seedling performance after planting (Binotto et al., 2010, Gomes et al., 2019).

Automatic classification of nursery seedlings based on morphological characteristics using computer vision technology has existed for some time. Machine vision technology is advantageous because it is objective, faster, and less error-prone than manual measurements (McGuinness et al., 2021). Computer vision technology uses cameras to obtain images of the seedlings. After that, the computer analyses the images through image processing and pattern recognition (Yan-Zhu et al., 2002). Seedlings can be graded and sorted using vision technology to improve their uniformity and quality. A prototype machine (laser profilometer) using point cloud processing software was found to be able to automatically measure seedling height in tree nurseries accurately (sub-centimetre) (Wanasinghe et al., 2019). In Brazil, a study evaluated the effectiveness of machine learning algorithms associated with hyperspectral data in predicting morphological variables and the quality of *Eucalyptus deglupta* seedlings in spectral variables (350–2,500 nm), with the aid of the hyperspectral spectroradiometer sensor, and morphological variables such as height, diameter, dry mass and relationships between these variables. The combination of machine learning algorithms with hyperspectral data proved to be efficient for predicting morphological variables and the quality of *Eucalyptus deglupta* seedlings (Ferriera, 2024).

Another study used binocular vision technology in conjunction with an advanced network structure to successfully measure seedling heights, achieving a success rate of 92.2% (Yuan et al., 2022). Programmed artificial intelligence vision systems can also detect plant stresses caused by diseases, pests and lack of water using red, green, and blue (RGB) thermal and spectral technologies (Mahmud et al., 2023). According to Rani et al. (2019), modern mobile devices with sensor technologies can be used to capture images, which are then checked against a database of existing images of diseased plants, enabling pathogen detection based on colour intensity, texture, spot count, and edge contour of the leaves.

Plant thermal health can also be assessed using computer vision and IoT to collect, analyse, and provide remedial processes such as temperature adjustments, air quality control, chemical release, light intensity management, and moisture control (Singh et al., 2020). Thermal technology using infrared and imagery analysis provides information on leaf temperature by analysing the stomatal responses of a specific species. Therefore, differences in stomatal conductance can be quantified to assess the behaviour of plants to water deficit (Jones et al., 2009). As the moisture in the soil is reduced, the leaf stomata close, resulting in reduced evaporation and increased leaf temperature. Based on temperature variances, the thermal imaging of nursery crops can be used to determine the water status and water stress level of plants and any changes in leaf temperature, which may be caused by other factors, such as diseases (Sharp, 2006). This technology in South Africa is still being investigated and tested for forest nursery applications (Naidu et al., 2022).

2.6.3 Track and tracing seedlings

An essential process in the nursery production process involves tracking and tracing seedlings. Drones can be used with trained models to identify and count seedlings within a specific area rapidly. Several research studies have found that detection and counting can be executed with high accuracy (Wang et al., 2023).

2.6.4 Robotics

A robot is an automatic device that executes tasks generally performed by a human (Bhavana and Bhagwan, 2021). In South African forest nurseries, no robots are currently being used to perform any operations. This technology may potentially be adopted in future as various technology drivers (e.g. cost of labour, labour shortages) influence the industry. Robotic technology is currently available in the horticulture industry to perform operations such as pruning, weeding, harvesting and disease control (Pekkeriet et al., 2015, Kutyrev et al., 2022, Sotnik and Lyashenko, 2022). Even though the cost is prohibitive for mass technology adoption, it is becoming cost-effective as refinements and efficiency improvements are implemented. The practical application of robotic systems will require integrating artificial intelligence, sensor-based solutions, digitalisation, and machine learning and deep learning. For example, robotic devices comprising a mobile system carrying a navigator, sensor, and detection and capturing technologies can be used to move along rows automatically sampling and conducting disease biosensing (Rani et al., 2019). Some companies, such as Flier Systems and ISO group, are investing in forest nursery robotic equipment, including the SPH-Transplanter (FlierSystem, 2024) and the ISO cutting Planter (ISO, 2023), respectively.

2.7 Ergonomics in nurseries

South Africa has very little literature on ergonomic studies and improvements in forest nurseries. Nursery work can be labour-intensive because some activities involve bending, lifting heavy containers, and exposure to chemicals, fertiliser, pesticides, dust and plant material (e.g. pollen). All these factors can negatively influence people's health, safety, and overall well-being (Posadas et al., 2004). The working postures of labour at the nursery can be hazardous.

Research findings have found that nursery workers can be affected by lower back pain, upper back pain, neck pain, shoulder and wrist pain (Yayli and ÇaliŞkan, 2019). In Brazil, ergonomic studies in nurseries focusing on repetitive tasks have found that activities such as tube filling, seeding, and transplanting are among the most likely to cause work-related musculoskeletal disorders (WMSDs), primarily when performed on benches at inadequate heights or without regular breaks (da Silva Lopes and de Oliveira, 2011). Adapted tools and layout reorganisation significantly reduced task completion time and fatigue rates among workers (Lopes and Oliveira, 2011, Possebom et al., 2017). Continuous ergonomic training and the adoption of preventive practices, such as stretching and job rotation, contribute to improved occupational health and productivity (da Silva Lopes and de Oliveira, 2011).

Ergonomically unfriendly work can harm productivity and overall employee wellness. Working methods and conditions can be improved to reduce the prevalence of musculoskeletal disorders (MSD). Applying mechanised systems to replace conventional labour-intensive methods in nurseries can reduce occupational workplace incidents (Varol et al., 2019) and has been a vital driver in overcoming numerous ergonomic challenges. According to da Costa (2024), in clonal nurseries, there has been a concerted focus on improving nursery designs to improve the ergonomics of nursery work by implementing efficient workflow/process flow and reducing distances travelled between different nursery production sites. For example, some nurseries have invested in sand bed technology, which is ergonomically designed for workers to work at the correct height of 900 mm with a reach of not more than 1.2 m. Additionally, in the setting shed, adjustments in work postures, sitting positions, and table designs are made to reduce the potential for MSDs. Work rotation to avoid repetition and the integration of work exercises have also been adopted in nurseries to relieve aches, pain, and stiffness. When carrying and moving loads, containers/trays can efficiently be moved by using roller tables instead of carrying individual trays from one position to another (da Costa, 2024).

2.8 Plant requirements for mechanised planting

When plants are delivered to the field for planting, the planting team needs to ensure that the root plugs are not damaged or deformed during the handling and planting process (Nelson, 2012). For details on different planting technologies, refer to Chapter 3. Advanced planting machines are designed to use containerised planting stock because the root plug is consistent compared to bare-root plants (Nieuwenhuis and Egan, 2002). For mechanised plants to succeed, the whole value chain of seedling production at the nursery and delivery at the site should be developed to suit mechanised planting operations (Rantala and Laine, 2010).

Before the seedlings are planted, they must be well watered, and the root plug must be consolidated. Planters are responsible for conducting quality checks and sorting the plants if they show signs of deformity, damage, or disease. In manual direct planting, the plant is carefully handled by hand until it is inserted into the prepared soil surface. In contrast, with planting tubes, the plant is gently deposited down a planting tube. However, in mechanised planting, the plant needs to be packed in a carousel at the head of a planting machine. The movement (gravity-fed in some cases) of the planting from this point will vary depending on the machine; however, the general principle is that the plant has to be dropped or slide from that point until it is planted in the soil (Ramantswana, 2023). Adherence to plant specifications and plant uniformity is critical for semi-mechanised and mechanised planting. The conditions are harsher because the seedling has to travel through planting tubes before being deposited into the ground during semi-mechanised planting. Therefore, the likelihood of mortality is high if the seedlings are soft (tiny, tender shoots, root plug not appropriately consolidated). Contractors involved in mechanised planting can experience seedling quality problems such as overgrown seedlings, undersized seedlings, loose root plugs, intertwined root systems, and diseased plants (Laine et al., 2016, Zwane, 2023). According to Biggs (2020), the plant specifications for effective mechanised planting are:

- Root plug consolidation (quality and integrity) by using paper pots enables the establishment of a robust root system, which captures the planting soil area more easily
- Sorting of plants that match the specification of mechanised planting (e.g. no double or multiple leaders)
- Uniform plant height and branching of plants produced from the nursery
- Ensure the timing of releasing plants to the field is correct
- Uniform root collar diameter and sturdiness
- Overall good plant health (no presence of disease or pest damage on the growing tip or stem)

2.9 Summary

This chapter explored nursery technologies used for regenerating fast-growing plantations in South Africa and Brazil. It detailed the processes for producing plants from predominantly cuttings, particularly in containerised nurseries. This chapter covered the core phases of plant production, including rooting, growing, hardening off, and dispatch, while emphasising the importance of watering, fertigation, and weed control. Plant quality assessment, emphasising seedlings suited for mechanised planting, was discussed. Furthermore, existing and emerging precision technologies, such as automated systems for climate control, irrigation, and plant health monitoring, were described. This chapter also addressed ergonomics in nurseries and the plant requirements for mechanised planting.

Review questions

1 Describe the application of containers in plant propagation at the nursery
2 With a focus on technologies used, explain what happens at the rooting stage of plant production at the nursery
3 With a focus on technologies used, explain what happens at the growing stage of plant production at the nursery
4 With a focus on the technologies used, describe the hardening-off stage of plant production at the nursery
5 Provide examples of automation technologies which can be used in forest nurseries

References

Almeida, D. 2016. Produção de sementes e mudas florestais. *Recuperação ambiental da Mata Atlântica.* 3ed. Ilhéus: Editus.

Barrett, G., Alexander, P., Robinson, J. & Bragg, N. 2016. Achieving environmentally sustainable growing media for soilless plant cultivation systems – A review. *Scientia horticulturae,* 212, 220–234.

Bhavana, H. & Bhagwan, A. 2021. Review on: Role of robotics in horticulture. *Journal of Pharmacognosy and Phytochemistry,* 10, 306–309.

Biggs, S. 2020. Paper pots as a solution for improved mechanised planting. *In:* Ramantswana, M. (ed.) Modern Silviculture Webinar, 23 October 2020 [Online]. George: Nelson Mandela University. Available: https://www.youtube.com/watch?v=ho_7uatdESI.

Binotto, A. F., Lúcio, A. D. C. & Lopes, S. J. 2010. Correlations between growth variables and the Dickson quality index in forest seedlings. *Cerne,* 16, 457–464.

Borges, S. R., Xavier, A., Oliveira, L. S. D., Melo, L. A. D. & Rosado, A. M. 2011. Enraizamento de miniestacas de clones híbridos de Eucalyptus globulus. *Revista Árvore,* 35, 425–434.

Brondani, G. E., Wendling, I., Grossi, F., Dutra, L. F. & Araujo, M. A. 2010. Miniestaquia de *Eucalyptus benthamii* × *Eucalyptus dunnii*:(II) sobrevivência e enraizamento de miniestacas em função das coletas e estações do ano. *Ciência Florestal,* 20, 453–465.

Cury, L. D., da Silva, L. F., de Oliveira, V. C., de Mello Arthur, M., Rodrigues, S. A., DA Silva, J. M. S. & Guerra, S. P. S. 2024. Crescimento de mudas de Eucalyptus urophylla produzidas em bandejas coletoras de água e convencional nos sistemas de tubetes degradáveis e polipropileno. *Ciência Florestal,* 34, e73734–e73734.

da Costa, D. 23 February 2024. *Interview. Advancements in silviculture technologies in South Africa.* Personal Communication.

da Silva, R. B. G. 2023. *Manejo de irrigação em viveiros florestais: uso racional da água e análise da qualidade*. São Paulo: Editora Científica Digital.

da Silva Lopes, E. and de Oliveira, F.M., 2011. *Evaluation of the demanded physical effort and posture of workers in forest nursery activity*. Cerne, 17(4), pp.573–582.

de Assis, T. F., Fett- Neto, A. G. & Alfenas, A. C. 2004. Current techniques and prospects for the clonal propagation of hardwoods with emphasis on Eucalyptus. *Plantation Forest Biotechnology for the 21st Century*. Trivandrum: Research Signpost, 303–333.

Delgado, L. G. M., DA Silva, R. B. G., Gabira, M. M., Rodrigues, A. L., Simões, D., DE Almeida, L. F. R. & DA Silva, M. R. 2022. Mean leaf angles affect irrigation efficiency and physiological responses of tropical species seedling. *Forests,* 13, 832.

Donald, D. 1986. South African nursery practice—The state of the art. *South African Forestry Journal,* 139, 36–47.

Dumroese, R. K., Luna, T. & Landis, T. D. 2009. *Nursery Manual for Native Plants: A Guide for Tribal Nurseries*. Washington, D.C.: US Department of Agriculture, Forest Service.

Durło, G., Jagiello- Lenczuk, K., Kormanek, M., Malek, S. & Banach, J. 2018a. Supplementary irrigation at container nursery. *Leśne Prace Badawcze / Forest Research Papers,* 79, 13–21.

Durło, G., Kormanek, M., Małek, S. & Banach, J. 2018b. Visualization and quantification of peat substrate moisture by fully automated moisture controlling system (SMCS) in forest container nursery. *Leśne Prace Badawcze / Forest Research Papers,* 79 (4), 319–327.

Ellepot. 2024. *Forestry 12+ ellepot paper* [Online]. Esbjerg: Ellepot. Available: https://www.ellepot.com/ellepot-products/paper/paper-programme/forestry-12plus/ [Accessed 16 February 2023].

Evans, J. & Turnbull, J. W. 2004. *Plantation Forestry in the Tropics: The Role, Silviculture, and Use of Planted Forests for Industrial, Social, Environmental, and Agroforestry Purposes*. Oxford: Oxford University Press.

Faustino, I. S., Jacovine, L. A. G., DE Morais, V. T. M., Moura, E. B. B., DA Rocha, S. J. S. S., DE Jesus França, L. C., Oliveira, K., Oliveira, T. A., Romero, F. M. B. & DE Paiva, H. N. 2022. *Avaliação da qualidade de mudas de espécies florestais nativas*. São Paulo: Editora Científica Digital.

Ferriera, F. A. R. 2024. *Dados hiperspectrais e aprendizado de máquina na avaliação do crescimento e qualidade de mudas de eucalipto.* Master in Agronomy, Federal University of Mato Grosso do Sul.

Fliersystem. 2024. *Our Solutions for Trees* [Online]. Nederland: Flier Systems. Available: https://fliersystems.com/en/cultivation-domains/trees [Accessed 2 March 2024].

Ford, C. 2013. Improving field survival of pine seedlings and cuttings: The Sappi plant quality Index©. *Proceedings of the International Plant Propagators Society-2013,* 1055, 11–16.

Garrido, M., Perez-Ruiz, M., Valero, C., Gliever, C. J., Hanson, B. D. & Slaughter, D. C. 2014. Active optical sensors for tree stem detection and classification in nurseries. *Sensors,* 14, 10783–10803.

Girard, D. & Gagnon, J. 2016. Wireless network of electronic scales to monitor the substrate volumetric water content for managing irrigation of containerized seedlings produced in forest nurseries. *Tree Plant. Notes,* 59, 9–19.

Gomes, S. H. M., Gonçalves, F. B., Ferreira, R. A., Pereira, F. R. M. & de Jesus Ribeiro, M. M. 2019. Avaliação dos parâmetros morfológicos da qualidade de mudas de Paubrasilia echinata (pau-brasil) em viveiro florestal. *Scientia Plena,* 1–15.

Grossnickle, S. C. & El-Kassaby, Y. A. 2016. Bareroot versus container stocktypes: A performance comparison. *New Forests,* 47, 1–51.

Grossnickle, S. C. & Ivetić, V. 2022. Root system development and field establishment: Effect of seedling quality. *New Forests,* 53, 1021–1067.

Grossnickle, S. C. & Macdonald, J. E. 2018. Seedling quality: History, application, and plant attributes. *Forests,* 9, 283.

Harders, L. O., Ufer, T., Wrede, A. & Hussmann, S. 2023. UAV-based real-time weed detection in horticulture using edge processing. *Journal of Electronic Imaging,* 32, 052405–052405.

Hechter, U., Little, K., Chan, J., Crous, J. & DA Costa, D. 2022. Factors affecting eucalypt survival in South African plantation forestry. *Southern Forests: A Journal of Forest Science,* 84, 253–270.

Hermisan. 2024. *SupraControl.* Alicante: Hermisan.

Idris, M., Salifu, K. & Timmer, V. 2004. Root plug effects on early growth and nutrition of container black spruce seedlings. *Forest Ecology and Management,* 195, 399–408.

ISO. 2023. *ISO Cutting Planter 2500* [Online]. Gameren: ISO Horti innovations. Available: https://www.tta-iso.com/equipment/cuttingedge-single [Accessed 19 February 2024].

Jones, H. G., Serraj, R., Loveys, B. R., Xiong, L., Wheaton, A. & Price, A. H. 2009. Thermal infrared imaging of crop canopies for the remote diagnosis and quantification of plant responses to water stress in the field. *Functional Plant Biology,* 36, 978–989.

Jooste, A. (ed.) 2022. *Tree Farming Guidelines.* Cascades: Sappi.

Kaptein, N. D., Savage, M. J. & Light, M. E. 2019. An irrigation control system with a web-based interface for the management of Eucalyptus planting stock in a nursery. *Southern Forests: A Journal of Forest Science,* 81, 31–37.

Kruger, M. 2012. Nursery practice in hardwoods. *In:* Bv, B. & Upfold, S. (eds.) *South African Forestry Handbook.* 5th ed. Pietermaritzburg: South African Institute of Forestry, 97–106.

Kutyrev, A., Kiktev, N., Jewiarz, M., Khort, D., Smirnov, I., Zubina, V., Hutsol, T., Tomasik, M. & Biliuk, M. 2022. Robotic platform for horticulture: Assessment methodology and increasing the level of autonomy. *Sensors,* 22, 8901.

Laine, T., Kärhä, K. & Hynönen, A. 2016. A survey of the Finnish mechanized tree-planting industry in 2013 and its success factors. *Silva Fennica,* 50, 1–4.

Land Jr, S. & Cunningham, M. 1994. Rooted cutting macropropagation of hardwoods. *Proceedings of the Southern Regional Information Exchange Group Biennial Symposium on Forest Genetics: "Applications of Vegetative Propagation in Forestry".* US Department of Agriculture, Forest Service, Southern Forest Experiment Station. Huntsville, Alabama, 75–96.

Landis, T., Dumroese, R. & Haase, D. 2010. Seedling processing, storage, and outplanting, Vol. 7, The container tree nursery manual. *Agricultural Handbook,* 674, 1–194.

Landis, T. D., Bonner, F. & Karrfalt, R. 2008. Nursery practices. *The Woody Plant Seed Manual,* 727, 125–146.

Landis, T. D., Jacobs, D. F., Wilkinson, K. M. & Luna, T. 1990. Growing media. *The Container Tree Nursery Manual,* 2, 41–85.

Landis, T. D. & Nisley, R. G. 1990. *The Container Tree Nursery Manual.* Washington, D.C.: US Department of Agriculture, Forest Service.

Liegel, L. H. 1987. *A Technical Guide for Forest Nursery Management in the Caribbean and Latin America.* New Orleans, LA: US Department of Agriculture, Forest Service, Southern Forest Experiment Station.

Lopes, E. D., Amaral, C. L. F. & de Novaes, A. B. 2016. Parâmetros morfofisiológicos na avaliação da qualidade de mudas de três espécies florestais. *Revista Agrogeoambiental,* 8, 1–8.

Lopes, E. D. S. & Oliveira, F. M. D. 2011. Avaliação da carga de trabalho físico e da postura de trabalhadores na produção de mudas florestais. *Cerne,* 17, 573–582.

Mafia, R. G., Alfenas, A. C. & Resende Júnior, M. F. R. D. 2007. Damping-off of forest species caused by Sclerotium rolfsii Sacc. *Revista Árvore,* 31, 629–634.

Mahmud, M. S., Zahid, A. & Das, A. K. 2023. Sensing and automation technologies for ornamental nursery crop production: Current status and future prospects. *Sensors,* 23, 1818.

Maile, N. & Nieuwenhuis, M. 1996. Vegetative propagation of Eucalyptus nitens using stem cuttings. *South African Forestry Journal,* 175, 29–34.

Mcguinness, B., Duke, M., Au, C. K. & Lim, S. H. 2021. Measuring radiata pine seedling morphological features using a machine vision system. *Computers and Electronics in Agriculture,* 189, 106355.

Mitshell, R. 2012. Pine propagation and nursery practice. *In:* Bv, B. & Upfold, S. (eds.) *South African Forestry Handbook.* 5th ed. Pietermaritzburg: South African Institute of Forestry, 89–95.

Mohanan, C. & Sharma, J. 2005. Improvement of seedling production system in forestry sector and its impact on seedling health. *Kerala for Research Institute,* 11, 77–82.

Naidoo, N. 2012. *Genetic and environmental factors affecting rooting in Eucalyptus Grandis X Eucalyptus Longistrata hybrid cuttings.* Master's thesis in Plant Breeding, University of KwaZulu-Natal, Pietermaritzburg.

Naidu, D., Peerbhay, K. & Brink, W. 2022. *Technologies for Improved Nursery Survival and Growth* [Online]. George: Nelson Mandela University. Available: https://www.youtube.com/watch?v=ho_7uatdESI [Accessed 28 February 2024 2023].

Naidu, R. & Jones, N. 2009. The effect of cutting length on the rooting and growth of subtropical Eucalyptus hybrid clones in South Africa. *Southern Forests,* 71, 297–301.

Nelson, W. 2012. Containerised Forest Nurseries. Christchurch: 888 Management Ltd.

Nieuwenhuis, M. & Egan, D. 2002. An evaluation and comparison of mechanised and manual tree planting on afforestation and reforestation sites in Ireland. *International Journal of Forest Engineering,* 13, 11–23.

Okorski, A., Pszczółkowska, A., Oszako, T. & Nowakowska, J. A. 2015. Current possibilities and prospects of using fungicides in forestry. *Leśne Prace Badawcze / Forest Research Papers,* 76 (2), 140–152. https://doi.org/10.1515/frp-2015-0014

Pekkeriet, E., VAN Henten, E. & Campen, J. 2015. Contribution of innovative technologies to new developments in horticulture. *Acta Hortic,* 1099, 45–54.

Poon, W. H. 1996. *A Machine Vision and Handling System for Measurement of Containerized Tree Seedlings.* Vancouver, BC: University of British Columbia.

Posadas, B., Fain, G., Coker, C., Knight, P., Veal, C. & Coker, R. 2004. Socioeconomic survey of nursery automation. *Proceedings of the Southern Nursery Association Research Conference,* Atlanta, GA, 306–309.

Possebom, G., Moreira, A. R., Carpes, D. P., Francetto, T. R., Zart, B. C. C. R., Do Amaral Alonço, P. & dos Santos Alonço, A. 2017. Avaliação ergonômica em um viveiro florestal de Santa Maria, RS. *Tecno-lógica,* 21, 30–36.

Poteri, M., Lilja, A. & Petäistö, R. 2005. Control of nursery diseases and pests in Finnish forest tree nurseries. *Working Paper of the Finnish Forest Research Institute,* 11, 19–26.

Rajković, S., Tabaković- Tošić, M. & Golubović- Ćurguz, V. 2007. Integral protection of forests – scientific-technological platform. *International Scientific Conference,* Belgrade, Serbia. Institute of Forestry, 83.

Ramantswana, M. 2023. *Plant Quality Is Key to Mechanised Planting*. George: Forsilvitech.

Rani, A., Donovan, N. & Mantri, N. 2019. The future of plant pathogen diagnostics in a nursery production system. *Biosensors and Bioelectronics,* 145, 111631.

Rantala, J. & Laine, T. 2010. Productivity of the M-Planter tree-planting device in practice. *Silva Fennica,* 44, 859–869.

Rigney, M. P. 1986. *Machine Vision for the Grading of Pine Seedlings*. Stillwater: Oklahoma State University.

Rihn, A. L., Velandia, M., Warner, L. A., Fulcher, A., Schexnayder, S. & Lebude, A. 2023. Factors correlated with the propensity to use automation and mechanization by the US nursery industry. *Agribusiness,* 39, 110–130.

Riikonen, J. & Luoranen, J. 2018. Seedling production and the field performance of seedlings. *Forest Ecology and Management*, 430, 740–748.

Schuermans, J. 2011. *The development of a wall-less plug for planting stock of forest trees.* Master of Science in Forestry, University of KwaZulu-Natal.

Shamshiri, R. & Ismail, W. I. W. 2013. A review of greenhouse climate control and automation systems in tropical regions. *Journal of Agricultural Science and Applications,* 2, 176–183.

Sharp, R. 2006. The application of advanced irrigation technology in Hardy Ornamental Nursery Bed Systems©. *Combined Proceedings International Plant Propagators' Society*, 194.

Shepherd, K. 1986a. *Plantation Silviculture*. Dordrecht: Martinus Nijhoff Publishers.

Shepherd, K. 1986b. Plantations and land-use. *Plantation Silviculture,* 22, 109–132.

Silva, A. C. R. D., Kulmann, M. S. D. S., Balieiro, F. D. C., Arthur, J. C., Zonta, E., Ribeiro, J. C., Benites, V. D. M. & Pereira, M. G. 2025. Biodegradable containers affect the morphological and nutritional aspects of Eucalyptus urophylla seedlings. *Revista Brasileira de Ciência do Solo,* 49, e0240216.

Singh, R., Singh, P. & Kharb, L. 2020. Smart nursery with health monitoring system through integration of IoT and machine learning. *In:* Tanwar, P., Jain, V., Liu, C.-M. & Goyal, V. (eds.) *Big Data Analytics and Intelligence: A Perspective for Health Care.* Emerald Publishing Limited, 117–136.

Sotnik, S. & Lyashenko, V. 2022. Agricultural robotic platforms. *International Journal of Engineering and Information Systems (IJEAIS),* 6 (4), 14–21.

Stape, J. L., Gonçalves, J. L. M. & Gonçalves, A. N. 2001. Relationships between nursery practices and field performance for Eucalyptus plantations in Brazil. *New Forests,* 22, 19–41.

Stuepp, C. A., Kratz, D., Gabira, M. M. & Wendling, I. 2020. Sobrevivência e crescimento inicial no campo de mudas de eucalipto produzidas em diferentes substratos. *Pesquisa Agropecuária Brasileira,* 55, e01587.

Tousignant, D., Lamhamedi, M. S., Colas, F., Rioux, M., Lemay, P. & Robert, N. 2007. New technological developments in cutting propagation to increase forest productivity in Quebec. *7e Carrefour de la recherche forestière–La connaissance éloigne les préjugés,* 19–20. Available: https://www.researchgate.net/profile/Fabienne-Colas/publication/242110234_New_technological_developments_in_cutting_propagation_to_increase_forest_productivity_in_Quebec/links/55c4c28d08aeb9756741e7e6/New-technological-developments-in-cutting-propagation-to-increase-forest-productivity-in-Quebec.pdf

Urbinati. 2024. *Tray filler* [Online]. Italy: Urbinati Nursery Technology. Available: https://www.urbinati.com/en/product/soil-tray-filler-rn12/#:~:text=The%20RN12%20filler%20is%20suitable,up%20to%20220%20mm%20high. [Accessed 16 February 2024].

Urbinati Nursery Technology. 2025. *Drum Seeding Line* [Online]. Available: https://www.urbinati.com/en/product/seeding-line-alfa65/[Accessed 10 December 2025].

Van Iersel, M. W., Chappell, M. & Lea- Cox, J. D. 2013. Sensors for improved efficiency of irrigation in greenhouse and nursery production. *HortTechnology,* 23, 735–746.

Van Wyk, G. & Verryn, S. 2000. Basic principles of tree breeding in South Africa. *In:* Owen, D. (ed.) *South African Forestry Handbook 2000.* Menlo Park: South African Institute of Forestry, 61–68.

Varol, T., Özel, H. B., Tuna, E. & Özdil, R. 2019. Efficiency comparison of mechanization techniques in nursery. *European Journal of Forest Engineering,* 5, 68–76.

Vasic, V., Konstantinovic, B. & Orlovic, S. 2012. Weeds in forestry and possibilities of their control. *Weed Control.* IntechOpen.

Wanasinghe, T. R., Dowden, B. R., DE Silva, O., Mann, G. K. & Lundrigan, C. 2019. Automated seedling height assessment for tree nurseries using point cloud processing. *International Conference on Robotics and Automation (ICRA).* Montreal: IEEE, 3686–3691.

Wang, A., Zhang, W. & Wei, X. 2019. A review on weed detection using ground-based machine vision and image processing techniques. *Computers and Electronics in Agriculture,* 158, 226–240.

Wang, Y., Sun, J., Wang, F. & Li, D. 2023. Remote sensing-based estimation of seedling density in nursery gardens using YOLOv4 deep learning algorithm. *Traitement du Signal,* 40, 1533–1541.

Williams, L. 2000. Nursery practices: Vegetative propagation. *In:* Owen, D. (ed.) *South African Forestry Handbook 2000.* Menlo Park: South African Institute of Forestry, 74–78.

Yan-Zhu, Y., Xue-Zeng, Z., Wei-Jie, W. & Xian, W. 2002. Review on the proceeding of automatic seedlings classification by computer vision. *Journal of Forestry Research,* 13, 245–249.

Yayli, D. & Çalişkan, E. 2019. A study on the health problems of forest nursery workers. *Bartın Orman Fakültesi Dergisi,* 21, 517–524.

Yuan, X., Li, D., Sun, P., Wang, G. & Ma, Y. 2022. Real-time counting and height measurement of nursery seedlings based on Ghostnet–YoloV4 network and binocular vision technology. *Forests,* 13, 1459.

Zwane, B. 2023. Early lessons on mechanized planting in Eswatini. *In:* Ramantswana, M., ed. *Modern Silviculture Webinar – Optimising Modern Silviculture* [Online]. George: Nelson Mandela University.

Zwolinski, J. & Bayley, A. 2001. Research on planting stock and forest regeneration in South Africa. *New Forests,* 22, 59–74.

3 Silviculture technologies in the establishment and re-establishment of forests

Muedanyi M. Ramantswana and Saulo P. S. Guerra

3.1 Introduction

This chapter describes silviculture technologies used for forest establishment or re-establishment. It will focus on key activities like residue management, planting position preparation, weeding, tree planting, and fertilisation. Furthermore, it will highlight how matching these technologies to specific stand requirements will be essential for enhancing plant survival, growth, and uniformity. The reader will learn about basic manual techniques and advanced mechanised methods used in South Africa and Brazil. Furthermore, various regeneration methods will be examined, focusing on both conventional and technological approaches, to provide foresters with insights into the available options and their applications in diverse forestry conditions. By understanding these advancements, foresters will be better equipped to optimise regeneration efforts and meet the demands of sustainable forestry practices.

3.2 Regeneration/establishment/re-establishment method

The successful regeneration, establishment, and growth of plantation forests depend on well-adapted genetic material and good silviculture throughout the rotation (Kanowski, 1997). The execution of good silviculture depends on understanding important silvicultural principles. In plantation forestry, various regeneration systems can be employed; in South Africa and Brazil, both natural and artificial regeneration systems are utilised (Zwolinski and Groenewald, 2004). Artificial regeneration is the predominant method used, involving the direct planting of seedlings (Nyland, 2016). This method is preferred because the landowner controls the planted species, establishment duration, and relative consistency in silvicultural treatments (Savill et al., 1997). Artificial regeneration activities include residue management, preparing a planting position, managing unwanted vegetation, tree planting, and fertilising. These regeneration activities are, by and large, labour-intensive, time-consuming and relatively costly (Evans and Turnbull, 2004).

DOI: 10.4324/9781003391227-3

3.3 Residue management

Site preparation is a silviculture term that embraces both residue management and preparation of a planting position. Site preparation involves an active intervention to enhance regeneration (Nyland, 2016, Smith et al., 1997). Residues refer to material generated during the stand tending and harvesting of trees. Residues include tree tops, branches, leaves, bark, cones, stumps, offcuts, and other plant debris (Achat et al., 2015). The quantity of harvest residues remaining on the site depends on the harvesting method applied (Ghaffariyan and Dupuis, 2021). The available technology and associated cost of managing the residues can be restrictive (Berry and Sessions, 2020). In South Africa, timber is harvested (annual volume: 16.4 million m^3) using different harvesting methods, including the cut-to-length (64%) method, tree-length (20%), and full-tree (16%) methods (Mathelele, 2019). In mechanised cut-to-length operations, harvesting residues are broadcast across the whole forest stand; in semi-mechanised operations, the residues are windrowed, stacked in piles or broadcast. Residues retained in the site can hamper mechanised soil preparation and subsequent stand-tending operations (Ramantswana et al., 2021) and significantly impact soil health, nutrient cycling, and overall plantation productivity (Rocha et al., 2016). The available technology and associated cost of managing the residues can be restrictive (Berry and Sessions, 2020). While residues can be utilised for bioenergy, generating additional revenue (São José et al., 2020), their long-term impacts on soil health and productivity should be considered (de Melo et al., 2022). Removing residues for energy use may be more economically advantageous in some cases, but this approach may not be sustainable in the long term (Rocha et al., 2016). In South Africa, timber is harvested (annual volume: 16.4 million m^3) using different harvesting methods, including the cut-to-length (64%) method, tree-length (20%), and full-tree (16%) methods (Mathelele, 2019). In Brazil, according to IBGE (2023), timber is harvested (with an annual volume of around 174 million m^3) using various mechanised harvesting methods, while firewood (with a yearly volume of around 60 million m^3) is still predominantly harvested manually.

In mechanised cut-to-length operations, harvesting residues are broadcast across the entire forest stand; in semi-mechanised operations, the residues are windrowed, stacked in piles, or broadcast. Residues retained on the site can hamper mechanised soil preparation and subsequent stand-tending operations (Ramantswana et al., 2021). Therefore, residues must be removed (e.g. burning, biomass collection), reduced (e.g. mulching), or cleared (e.g. v-blade).

In South Africa, harvest residue management methods include burning, spreading, mulching, biomass removal, windrowing, and using a soil preparation machine (Ramantswana et al., 2020). However, burning of harvesting residue or any biomass collection was prohibited in Brazil in 1998 with the enactment of Law No. 9,605, the Environmental Crimes Law.

3.3.1 Controlled burning

Controlled burning of residues, also known as slash pile burning, is the most prevalent method of residue management in South Africa (Ramantswana et al., 2020). Inter-rotational fires are used primarily to reduce fire hazards and debris on site after harvesting (Evans and Turnbull, 2004, Nadel et al., 2007). Burning of harvest residues can increase the availability of nutrients during the early stages of tree growth; however, it also increases the risk of soil erosion and machinery impacts on the soil (Titshall et al., 2013). When conducting controlled burning, the aim is to achieve a cool burn and avoid high-intensity fires that detrimentally affect the soil's organic material (Figure 3.1). A low-severity burn leaves the forest floor intact; however, a severe burn removes material from the forest floor and discolours the surface (Bigley and Henderson, 1989). Most residue burning in South Africa is limited to the summer season, when conditions (moisture and weather) are more suitable for controlled fires.

Harvest residue can be presented in various forms, depending on the harvesting method and system, which in turn affects the burning approach. For example, a mechanised cut-to-length system, comprising a harvester and forwarder, will leave the residues broadcast across the entire stand. Still, a mechanised full-tree system (Feller bunch, grapple skidder and roadside processor) will concentrate the residues along the roadside where all the processing occurs. A tree-length harvesting system will have tops and branches across the site, while the offcuts derived from optimising timber into different assortments will be concentrated along the roadside. In motor manual harvesting operations, residues can be left in windrows or piles for more concentrated burning. The piles for burning can be built using an excavator with a grapple or a three-wheeled machine to collect fine and larger woody materials, creating piles or windrows.

Figure 3.1 Residue burning

Burning is conducted manually by a team of trained firefighters, who ignite, monitor, and keep the fire within the designated stand area. Where the residues are broadcast, the fire burns sparsely across the stand, depending on the fuel concentration. Where residues are concentrated, the fire burns hot in a small area, affecting the soil (e.g. nutrients, infiltration, and structure) (Page-Dumroese et al., 2017). Equipment such as purpose-built fire trucks and tractor-trailer vehicles carrying water supports the burning operation (Refer to Chapter 4 for Forest protection technologies). The firefighters carry backpacks with water to extinguish any fires outside the designated zone. Burning is a high-risk activity; therefore, current and future weather conditions are considered to ensure that the burning operation is conducted safely and in an environmentally friendly manner.

3.3.2 Mulching and stump clipping

Mulchers or masticators are responsible for cutting, chopping, or grinding residues or vegetation into smaller pieces that are deposited at the site (Halbrook et al., 2006). Mulching involves spreading a specific material evenly over the soil surface (Solgi et al., 2021). Mulch from forest residues can cover the soil, prevent soil and nutrient wash-off, retain moisture and nutrients, and suppress weed growth (Prats et al., 2023, Chapman, 2020). In South Africa, mulching was found to be the third (8.4% of the total surveyed area of 41,869 ha) most applied method of residue management (Ramantswana et al., 2020). The application of this method has been increasing not only for harvesting residue management but also for fire protection and fuel load reduction (Chapman, 2020). Mulchers can be used to mulch different types of harvesting residues in preparation for soil preparation and planting. Mulchers can mulch broadcasted, windrowed, and moderately piled residues irrespective of the harvesting method.

Mulching equipment can have a traction system comprising wheels or tracks. Carriers can be purpose-built for forestry or adapted from agricultural equipment (e.g. tractors) (USDA Forest Service, 2004). Wheeled mulchers are more versatile because they are mobile, quicker and can easily overcome obstacles on the ground compared to tracked mulchers (Figure 3.2). However, wheeled machines are limited to gentle and moderate terrain, but tracked machines can work on steeper terrain. Two types of mulching attachments are used in forestry: drum and disc mulchers. Drum mulchers are commonly used to grind down hard materials, such as stumps, branches, and treetops, simultaneously. In contrast, disc (rotary) mulchers quickly cut and feed materials like trees and branches (Swift, 2020). In South African forestry, horizontal shaft mulchers with continuously spinning drums, as the machine moves through the material, are common.

The use of a purpose-built wheeled mulcher was found to be more productive (1.5 h ha^{-1} – two passes) and yield better work quality (>70% fragments

Figure 3.2 Purpose-built wheeled mulcher

shorter than 20 cm) compared to a non-purpose-built tractor mulcher (3.6 h ha^{-1} and <30% shorter than 20 cm) when operating in harvested pine sawtimber stands (Zitha et al., 2023). Furthermore, it was found that there was no significant difference between mulching eucalypt (0.38 ha PMH^{-1}) and pine (0.35 ha PMH^{-1}) stands (Mahadeo, 2022).

High stumps can hinder access for timber extraction and silviculture equipment after harvesting. Furthermore, high stumps can cause tyre punctures and damage the underbody of equipment that lacks adequate clearance. In South Africa, a locally designed stump clipping machine has been developed to reduce the high volume of stumps after harvesting. The stump clipping machine consists of an excavator equipped with a clipping attachment. The blades on the attachment sever the stump at ground level or slightly below ground level (Figure 3.3a). The clipper does not uproot the stumps from the ground, however, in Brazil the stump is uprooted completely from the ground (Figure 3.3b). The recovered stump wood material can be used for biomass and firewood or retained on site.

3.3.3 Windrowing

Windrowing is a site preparation process in which surface residue material is moved and accumulated in the form of a windrow (long parallel lines) manually or mechanically. In semi-mechanised harvesting operations (e.g. hardwood pulpwood harvesting), materials are windrowed manually and subsequently burnt once the harvesting is completed. The harvest residue material can also be mechanically moved using an excavator with a bucket, slash rakes, a dozer, or tractor-mounted shearing blades (Morris et al., 1983).

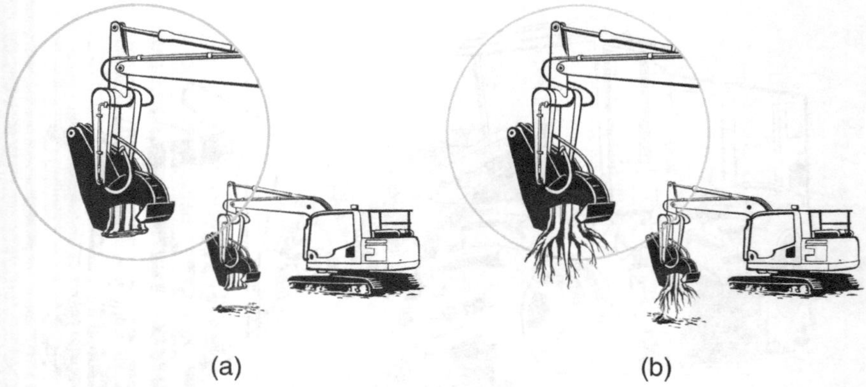

Figure 3.3 Stump clipper and stump uprooting

The material is gradually shifted and accumulated into a windrow, either burnt or left to decompose on the site. The windrowing activity is conducted to clear and enable access for soil preparation and planting operations (Clarke et al., 2015). Furthermore, if harvesting residue is collected for bioenergy purposes, windrowing can facilitate improved recovery and collection efficiency. Due to the dominance of burning for residue management in South Africa, mechanical windrowing is limited.

3.3.4 Chopper rolling

Chopper rolling is a residue management process in which a heavy drum with parallel metal blades is pulled over residues on a site to break and reduce the size of the material retained on the site (Sullivan et al., 2021). A chopper roller drum(s) (also referred to as a roller chopper) can be towed behind a tractor or dozer. The drum can be emptied or filled with water to increase its weight. Chopper rollers are expensive and are predominantly suited for gentle terrain conditions and areas with good ground roughness (Allan et al., 2000). This limitation prohibits their wider application. In South Africa, chopper rolling technology was prevalent from the early 1990s to the late 2000s (Ndlovu et al., 2019, Roland et al., 2008), but its use has steadily declined with the introduction of mulchers in forestry.

3.4 Marking a planting position

The establishment and re-establishment spacing influences total volume production and tree size. Some standard spacings in eucalypt pulpwood rotations are 3 × 2 m, 3 × 2.5 m, and 3.5 × 2.5 m. In sawtimber, spacings are typically 3 × 3 m or even wider. The marking of a planting position to achieve the required stand spacing is predominantly manual, involving the setting up of a baseline (71.6% of the total surveyed area, 37,778 ha)

Figure 3.4 Manual marking and pitting of a planting position

(Ramantswana et al., 2020). The manual method consists of using long ropes or wires with markers at required spacing intervals, attached to poles at both ends (Figure 3.4).

When marking a planting position, pitting can be performed immediately, or a spot can be marked with a hoe or stick for a separate pitting operation (manual, semi-mechanised, or mechanised) (Evans and Turnbull, 2004). Brazil has not adopted this methodology because most forest plantations will be established using subsoiled lines, which are tractor-guided with GPS. The marking can be done by tying a piece of rope to the base of the manual planter to determine the spacing between plants in the planting row, using the desired spacing. In re-establishment cases where the existing spacing will be retained with the new plant crop, no new marking is required as old stump lines (Ramantswana et al., 2020) are used as references. There is still a technology gap when it comes to accurately mapping the location of manually prepared pits; however, emerging technologies that utilise smartphones, smartwatches, or real-time global navigation satellite system (GNSS) to generate maps can be a solution for the future (Keefe et al., 2022).

With the increase in mechanised soil preparation, new technologies such as GPS systems or wheel encoders can be automatically integrated into soil preparation machines to perform pitting or ripping the soil using georeferencing systems. According to Nsele (2023), the use of digital tools to identify the area where a pit must be made requires:

- Planning: an online platform for compartment and workflow planning and optimisation is required. This task can be performed by using a website grid generator

- Operating: GPS operator guidance is needed to plan pit and planting locations and record actual coordinates to facilitate further silvicultural operations
- Reporting: a collection of data in the mechanised operations eliminates manual record keeping and produces on-demand production reporting of total pitting/plantings per compartment, shifts, and total time worked in a compartment, average pitting/plantings per hour, stems per hectare and actual GPS coordinates of each location

3.5 Preparation of a planting position

Preparation of a planting position involves breaking the soil after felling to facilitate the growth of a new generation of trees. The primary objective of preparing a planting position is to enhance anchorage to the soil through the root system, improve aeration, and facilitate access to water and nutrients (de Moraes Gonçalves et al., 2004). The preparation of planting position methods can be categorised into two broad categories: surface cultivation (pitting, rotovating, ridging, discing, or ploughing) and deep cultivation (ripping, subsoiling, or terracing) (Viero, 2004). Preparation of a planting position can be conducted manually or mechanically. Soil preparation can be performed as a separate activity before planting or integrated into the planting process. In some areas, no soil preparation is undertaken; a hole is opened at planting. Generally, the tree growth responses to specific soil preparation methods are often site-specific; hence, a generalised prescription for soil preparation can yield different results (Smith et al., 2001). According to Jooste (2022), the advantages of preparing a planting position are:

- Decrease in soil bulk density and physical strength
- Improve water infiltration
- Promotion of oxygen diffusion rates
- Increase the rate of organic matter decomposition in the topsoil
- Removal of weed competition around the seedling

A good and acceptable pit should be deep enough to allow the seedling root plug to be planted into the soil and should contain soil with a fine tilth (no undecomposed debris) (Light, 2015). The productivity of a manual, motor manual and mechanised preparation of planting position operations can be affected by the level of residues on the site (e.g. burnt or unburnt), terrain conditions (flat or steep), ground conditions and presence of obstacles such as stumps (Jacob, 2022, Rolando and Little, 2006). Soil preparation is a costly activity; therefore, the most effective method must be used without compromising soil fertility and negatively impacting tree growth (Alcázar et al., 2002). Even though preparing a planting position is critical, it is also essential to ensure that best practices are adhered to when conducting associated activities, such as planting and weed control (Light, 2015). This section will focus on the different pitting methods, pitting quality, and the effects of pitting on plant growth.

3.5.1 Manual pitting

The manual preparation of a planting position remains the most widely practised method (57.2% of 39,013 ha surveyed) of soil preparation in South Africa (Ramantswana et al., 2020). In Brazil, manual soil preparation accounts for less than 1.0% of the 2,796,669 hectares surveyed for eucalypt forest plantations. However, for pine forest plantations, 7.0% of 223,279 ha surveyed was prepared using manual tools (Guerra et al., 2023). Manual soil preparation is conducted by pitting, which creates a planting position by loosening the soil using a hand tool, such as a pick or mattock (Figure 3.5).

For re-establishment purposes, some growers work with pit dimensions of 25 cm deep and 35 cm wide, while for establishment, they use pits of 25 cm deep and 50 cm wide (Jooste, 2022). Even research suggests that a

Figure 3.5 Manual pitting

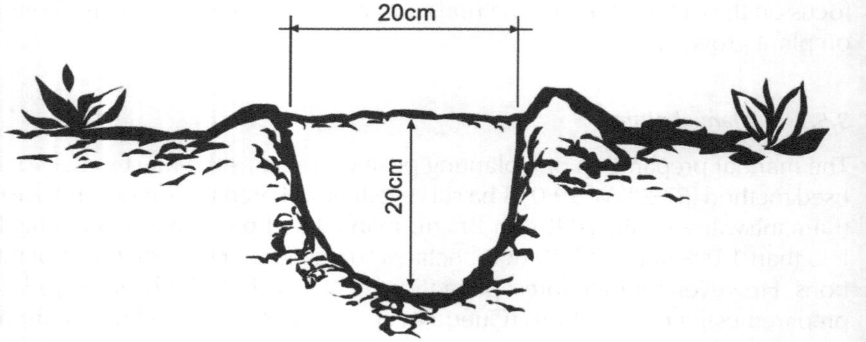

Figure 3.6 Pit dimensions

standard depth of 15–20 cm with a diameter of 20 cm is deemed adequate (Figure 3.6). Jacob (2022) found that manual pitting in unburnt and burnt eucalypt sites yielded work productivity rates of 108 and 174 pits per hour (spacing 3 × 2.5 m), respectively. In burnt and mulched pine sites, Mahadeo (2022) measured productivity of 101 and 87 pits per h^{-1} (spacing 3 × 2 m). Pit quality is measured by assessing the depth, width, friability, and volume of the pit. The depth and width of a pit depend on the pit preparation method applied (Hechter et al., 2020). Pit depth is the criterion directly affected by residues on a site. There is an overall tendency to have shallower pits when a slash is present on the site (Rolando and Little, 2006). Considering both burnt and unburnt sites, Jacob (2022) determined a mean pit depth of 21.9 cm, whereas Rolando and Little (2006) measured pit depths varying between 25 and 29 cm on sites with harvest residues broadcast, chopper rolled, and burnt.

3.5.2 Motor manual pitting

Soil preparation can be manually conducted using a motorised tool, such as an earth auger. The earth auger is a handheld machine manually carried to each location, powered up, and guided to drill holes in the ground (Figure 3.7). The motor manual preparation of a planting position is the fourth most practised method (5.3% of 39,013 ha surveyed) of soil preparation in South Africa (Ramantswana et al., 2020). In Brazil, motor manual pitting represents 3.0% of 2,796,669 ha surveyed for eucalypt forest plantations. For pine forest plantations, almost 7.0% of the 223,279 ha surveyed was prepared using motor manual tools (Guerra et al., 2023). An earth auger can be coupled to different attachments such as the Archimedes screw, soil loosening type (Inverted A bit) and Mondi designed (Hechter et al., 2020, Ackerman et al., 2018). According to Gonçalvez JLM (2002), for a preparation depth of 40 cm, the subsoiler mobilises a volume of soil of approximately 531 m^3 ha^{-1}, the mechanical auger a volume of 83 m^3 ha^{-1} and the manual pit 45 m^3 ha^{-1}. Residues also influence pit depth in motor manual pitting operations. The pit

Figure 3.7 Motor manual pitting

widths of auger-prepared pits are limited to the width of the pitting attachment. The attachment can also affect friability and pit soil retention; some attachments may disperse the soil away from the pit, while others retain the soil in and around the pit. An earth auger's pitting operation typically comprises two operators who alternate using the auger throughout the shift. In Brazil, only one operator uses the earth auger pitting for the 8-hour working shift. This operation is limited to steep slopes or areas with physical obstacles, such as rocks or tree stumps, where tractors or heavy equipment (subsoilers, rippers, or similar) cannot perform correctly. Jacob (2022) found that motor manual pitting in unburnt and burnt eucalypt sites yielded productivities of

314 and 219 pits per hour (3 × 2.5 m), and the mean pit depth was 21.9 cm (7.5% pit compliance). However, in a different study, Ramantswana (2023) reported a productivity of 184 pits per hour (3 × 2.5 m) in a burnt stand and found that 80.6% of the pits produced were compliant with a pit depth of 25 cm. From general observations, it can be inferred that pit quality is negatively affected by high productivity. Studies in eucalypt stands that have compared manual and motor manual pitting have not found significant differences in tree growth at 12 months (Jacob, 2022, Hechter et al., 2020).

3.5.3 Mechanised pitting

Mechanised pitting is described as using a machine to loosen the soil and form a pit for planting (Figure 3.8). In South Africa, mechanised pitting technologies have evolved (around 2007) from wheeled carriers attached with single- and multiple-auger pitting implements (Crane, 2017) to predominantly tracked excavator-based carriers with a single head (Chapman, 2012). This method has steadily increased over the years and has become the second (16.4% of 39,013 ha surveyed) most-used method of soil preparation in South Africa (Ramantswana et al., 2020). Mechanised pitting produces consistent depth (31 cm) and width (30 cm) compared to manual and motor manual pitting operations, even in areas with residues (Jacob, 2022). This method is more capital-intensive than manual-oriented methods. However, it can simultaneously produce between three and four times more pits (300–500 pits h^{-1}). The productivity of single-pitting head machines can be influenced by residue status, marking system, number of pitted lines, tree spacing, and operator experience (Mahadeo, 2022, Jacob, 2022, Zulu, 2023). With mechanised pitting producing more bottomless pits, there is a concern that plants may be planted too deeply. Preliminary results, limited to one area from Jacob (2022), found that the biomass index of eucalypt plants planted in mechanically prepared

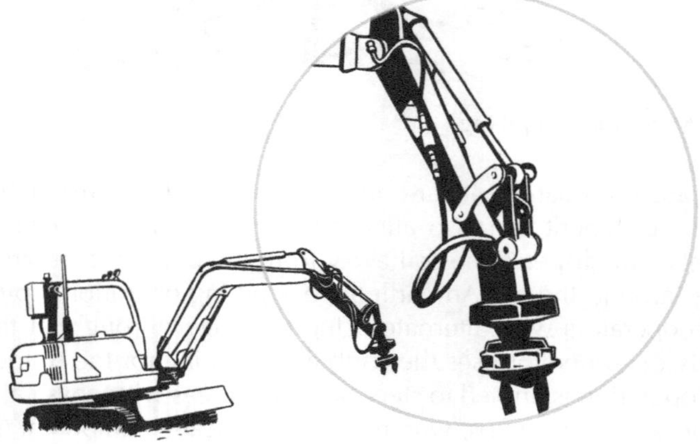

Figure 3.8 Single-pitting head machine

pits was significantly lower than that of plants planted in manually prepared pits after 12 months.

3.5.4 Subsoiling or ripping

Subsoiling and ripping are terms often used interchangeably (du Toit et al., 2010) to refer to a mechanical soil preparation method that opens a narrow trench either by loosening and or breaking hard subsoil layers using a metal tine (can be up to 1 m) (Smith et al., 2012, Evans and Turnbull, 2004). In South Africa or in Brazil, ripping generally refers to using tines to rip up to depths of 45 cm, and the lateral slope did not exceed an inclination of 40%–60% (Lima et al., 2004, Pereira et al., 2012) as preconised by Chudakov and Okúlik (1977) and Mialhe (1980). Subsoiling is typically associated with deeper ripping depths (greater than 40 cm) (Smith et al., 2001). In Brazilian forest plantations, subsoiling is a widespread method, representing almost 80% of the surveyed area in 2018 and more than 95% of the 2,796,669 ha surveyed for eucalypt forest plantations in 2022. For pine forest plantations, almost 86% of the 223,279 ha surveyed were prepared using manual motor tools (Guerra et al., 2023). Many studies show that more intensive soil preparation methods (such as subsoiling or ripping) can significantly lead to more remarkable tree growth and biomass production of both eucalypts and pines (Gatto et al., 2003, Paes et al., 2013, Reichert et al., 2021). The benefits of deep soil preparation are more pronounced in cohesive soils, where compaction restricts root penetration (Souza, 2002, Souza and Silva, 2001). This provides a better soil condition for tree development, allowing them to exploit a larger volume of soil and enhancing nutrient and water absorption. Subsoiling improves soil aeration, which aids in decomposing organic matter and litter, as well as harvesting debris, thereby releasing nutrients for plant uptake (Baptista, 2008). It also ensures uniform soil conditions and promotes even production across the stand (Fessel, 2003).

Ripping is performed using a ripping shank on a tractor, dozer or skidder (Figure 3.9). The power requirements of the tractor when ripping are influenced by the soil strength, ripping depth (shallow or deep), and residues on the site (Espinoza, 2004). Growth responses from ripping have been widely studied, with mixed responses related to the site and soil type (Smith et al., 2001). In South African conditions, it is understood that the reasons for limited growth are principally climate-related rather than soil-related (Smith et al., 2001). Even though Ndlovu et al. (2019) found that when establishing *P. patula*, the ripped area's basal area and rotation end volume were significantly better than the pitted, chopper rolled + pitted as well as chopper rolled + ripped areas; Smith et al. (2001) found that intensive site preparation techniques like ripping did not generally improve the growth of *E. grandis* and *A. mearnsii* on various sites in South Africa. According to Zwolinski et al. (1996), ripping can suppress weeds during the early growth phase; hence, growth gains may be derived from the reduction in

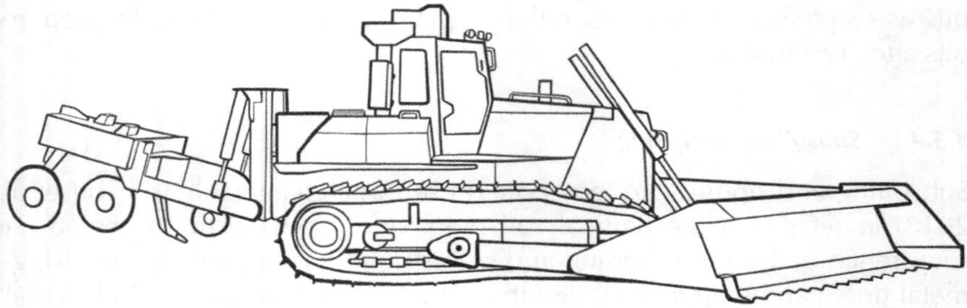

Figure 3.9 Dozer subsoiling machine

competition instead of soil amelioration effects. Soil preparation must be carefully planned to ensure minimum soil impact and reach the desired final productivity (Fessel, 2003). When ripping is not applied correctly and according to the soil type, mineral soil is disturbed and exposed. It is also essential to be cautious when ripping on slopes as it can facilitate soil erosion, increase water loss through deep drainage and negatively affect future site productivity due to potential water availability reduction for plants (Alcázar et al., 2002, Paes et al., 2013).

3.6 Weeding before planting

Weed control during the early establishment phase is a critical silvicultural practice to prevent weeds from competing with plants for water, light, and nutrients. Weeds can negatively impact plants by reducing their qualitative and quantitative characteristics, by serving as hosts for diseases and pests, through allelopathy, or by interfering with further irrigation, fertilisation, or silvicultural operations (Silva et al., 2012). When adhering to the clear-fell even-aged silvicultural system, harvested areas must be re-established as soon as possible to prevent weed infestation in the open stand. Weed control can be executed manually, chemically, mechanically or aerially (Lyle, 1981). The method used to eradicate weeds before planting depends on various factors, such as prevailing climatic conditions, types of weeds, intensity of the weed infestation, terrain, and costs. The lack of weed control before planting can inhibit early tree growth, which may not be recovered with remedial weed control and fertilising actions later in the early growth phase (Alwi et al., 2020). This section focuses on the implements and technologies used to conduct weed control before planting, while acknowledging the existence of manual methods, such as slashing. In South Africa, weeding before planting can also be incorporated into residue management activities such as burning and mulching, but not in Brazil. Although the primary goal of residue management activities is to remove, reduce, and/or redistribute residues, existing weeds can also be removed during this process.

3.6.1 Manual chemical spray

Manual chemical spray is a method that uses tools such as knapsack sprayers to apply chemicals to weeds (Figure 3.10). In South Africa, chemical spraying with a knapsack is the most common method (70.8% of the total 37,953 ha surveyed) for weed control before planting (Ramantswana et al., 2020). In Brazilian forest plantations, manual chemical spray is not the most-used method, representing 30.3% of the surveyed area of 2,796,669 ha for eucalypt forest plantations in 2022. For pine forest plantations, almost 43% of the 223,279 ha surveyed were controlled with manual spraying tools (Guerra et al., 2023). Chemical sprayers can be grouped according to the total volume of mixture applied per hectare. In re-establishment, weeding before planting involves a blanket spray when weeds are 150 mm (Lyle, 1981). A minimum of 75 cm of the treated strip is recommended to minimise weed interference with eucalypt plants´ growth (Silva et al., 2012). The operator must be trained to wear the correct protective equipment when using chemical spraying. Before applying any chemical, Salomão et al. (2020) suggested that the vegetation and weather conditions – temperature below 30°C, air moisture content higher than 60%, and wind speed up to 10 km h^{-1} (Bugatti et al., 2012) are more suitable for spraying, and the equipment must be calibrated to reduce spray drift. Manual chemical spraying is advantageous because of low capital costs, less complexity, and more flexibility. It can also be used in areas with poor access. However, manual chemical spray has low output rates, and operators can be exposed to dangerous chemicals and poor ergonomics when carrying a load on the back (Jones, 2006). Therefore, workers should be encouraged to take breaks during the workday for rest and muscle relaxation, mainly since the job is performed in a standing position (Lopes et al., 2011).

Three main types of knapsacks can be used for chemical spraying: hand-operated, motorised, and electric. The hand-operated knapsack sprayer is the conventional and most-used sprayer. It is preferred because

Figure 3.10 Manual chemical weeding before planting

it is cost-effective and relatively easy to repair and maintain. This type of sprayer has a low work efficiency and is labour-intensive (Guanfeng, 2022). Electric high-pressure automatic sprayers are more efficient and have better atomisation intensity. These sprayers are more costly to maintain, and pump repairs are more complex. The battery life of the sprayer limits the length of continuous work. The motorised knapsack comprises an engine-powered spraying system. A motorised knapsack sprayer can increase spraying efficiency and enhance precision; however, the purchase and operational costs are higher because they require fuel to operate (Damilare, 2023).

3.6.2 Tractor boom spray

Tractor boom spraying technology comprises a carrier vehicle coupled to a tank and boom sprayer with nozzles, enabling chemical application on a specific site. The tank can be mounted on the tractor's three-point linkage or towed as a trailer. The chemical can be used in weeding before the plant is fully covered, broadcast, or overall spraying, which means the application covers the whole area and targets all undesirable vegetation (Jones, 2006). The herbicide spraying process is activated by engaging the tractor's hydraulic or power take-off (PTO) drive system, which enables the herbicide to move through the piping to the boom, which is released through the nozzles (Roberts et al., 2021).

In South Africa, tractor boom spraying technology is the second most prevalent method (14.7% of the total 37,953 ha surveyed) used for weeding before planting (Ramantswana et al., 2020) (Figure 3.11). In Brazilian forest plantations, semi-mechanised and mechanised weed control is widespread, representing almost 66% of the 2,796,669 ha surveyed for eucalypt forest

Figure 3.11 Tractor boom sprayer

plantations in 2022. For pine forest plantations, approximately 50% of the 223,279 ha surveyed was prepared using semi-mechanised and mechanised weed control equipment (Guerra et al., 2023).

The Troner sprayer is a tractor boom spraying technology used in South Africa, comprising an adjustable boom (18 m wide with 36 nozzles), a 2,300-litre tank, and a separate 30-litre herbicide mixing container (Nsele, 2023). It utilises a computer and GPS that report variables such as water used, travel speed, and areas sprayed. Furthermore, the applicable nozzles will be deactivated if the spraying overlaps an already-sprayed area. The Troner tractor has been reported to achieve a productivity of 4.5 ha PMH^{-1} (row length 472 m) in comparison to a conventional tractor sprayer, which achieves between 1.5 and 2.3 ha PMH^{-1} (Nsele, 2023). A tractor with a boom sprayer can have a GPS receiver and an onboard computer to guide the tractor carrier in the stand during spraying. The GPS receiver must be correction-compatible to improve accuracy. The system can then be used to guide the operator and automatically steer the tractor to avoid re-spraying areas that have already been sprayed. The variable rate control system can also be integrated to apply the required quantities of chemicals when needed (Sciences, 2005). In agriculture, fully automated tractors can automatically spray herbicides with relative accuracy by using accurate GPS through real-time kinematic (RTK)-GNSS technology (Pérez-Ruiz et al., 2015). It can be expected that this type of technology will be adopted in forestry applications in the future.

3.6.3 Unmanned aerial vehicles

An unmanned aerial vehicle (UAV) is a powered aerial vehicle that does not carry a human operator and uses aerodynamic forces to provide vehicle lift. It can fly autonomously or be piloted remotely, is expendable or recoverable, and can carry a payload (Bone and Bolkcom, 2003). UAVs used for applying herbicides and pesticides have increased in number over the past decade (Figure 3.12). Even though UAVs are commonly used for remote sensing, nutrient assessment, plant health assessments (pest and disease detection), water stress analysis, yield and biomass estimates, soil monitoring, weed detection, and environmental monitoring, their application in aerial spraying is increasing (Del Cerro et al., 2021, Torresan et al., 2017, Medauar et al., 2018, Kim et al., 2019).

UAVs can be classified broadly according to performance specifications, including weight, endurance and range, maximum altitude, wing loading, engine type, and power/thrust loading (Arjomandi et al., 2006). Most UAVs are powered by electricity generated from battery power; however, some vehicles are powered by combustion engines (Amaral et al., 2021). There are different types of UAVs: fixed-wing, single-rotor, multirotor (tricopter, quadcopter, hexacopter, and octocopter) and hybrid vertical take-off and landing (VTOL) (Velusamy et al., 2021).

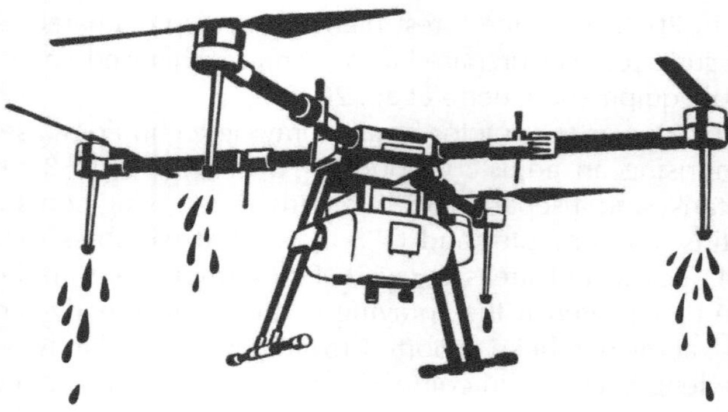

Figure 3.12 Unmanned aerial vehicle for spraying

The use of UAVs has several spraying advantages (Carvalho et al., 2020):

- With the use of GPS and sensors, the precise application of herbicides on a site can be executed
- More efficient spray operations (use of water and herbicide resources)
- More fuel efficient compared to conventional aircraft used for spraying
- Spraying can be carried out in multiple locations in a single day
- Rotors can increase the penetration and effectiveness of herbicides
- Due to the variable speed and nozzle locations, better herbicide penetration can be achieved compared to conventional aircraft
- No compaction compared to wheeled-based vehicles
- Better access to spray on rugged terrain (e.g. steep slopes, poor ground conditions, roughness, and inconsistency)

The challenges of using UAVs for spraying (Velusamy et al., 2021, Chapman, 2021):

- Costs of using UAVs can be high (cost of sensors, applications, and software for data analytics)
- Weather conditions such as rain, snowfall, clouds, and fog limit UAV activities
- Limited payloads
- Limited flight duration due to battery life – high auxiliary time (up to 66% of the time)
- Limited operational efficiency – from 20% to 35%
- Prevalence of obstacles (e.g. powerlines, fences, and trees along the flight path)

Even though the challenges mentioned above are prevalent, UAV technologies are constantly being upgraded (e.g. increasing payloads) to overcome the challenges (Wang et al., 2022). In both countries, the use of UAVs is still in the early phase, but it is expanding as contractors and grower companies adopt the technology. Very little scientific information is available on UAVs' productivity, costs, and work quality. Remotely piloted aircraft system (RPAS) drones are commonly used to apply glyphosate to eradicate weeds before planting. In Brazil, automatic weed control or pest control has become an early-stage reality, representing almost 4% of the 2,796,669 ha surveyed for eucalypt forest plantations in 2022. That is relevant information and a change in Brazilian mechanisation because, in 2018, UAVs were not shown working in the fields. For pine forest plantations, practically 10% of the 223,279 ha surveyed was prepared using automatic weed control, but still, no UAV technologies for pest control were applied (Guerra et al., 2023).

A pre-plant chemical operation with a drone starts with mapping the stand and setting the boundaries using a GPS. The software application can determine the most efficient way to spray the stand. A drone with a payload capacity of 10 litres of glyphosate chemical mix can spray 2–2.5 ha h^{-1} from an elevation of 2–3 m above ground (Chapman, 2021), reaching up to 2.3 ha per hour in Brazil (Programa Cooperativo sobre Mecanização e Automação Florestal, 2021). Support for battery replacement and chemical mix replenishment when the drone is empty is essential to ensure efficient operation. A larger drone, such as a 30-litre spray tank volume, can achieve 8–10 ha h^{-1} in fair weather conditions when spraying herbicides or pesticides (Greyling, 2022). The 40-litre spray tank volume, one of the latest drones, has an operating efficiency of 11–14 ha h^{-1}, including tank and battering changes (PACSys, 2024). Training and compliance with company and drone pilot licensing are a legislative requirement before any aerial applications are conducted.

In the context of ant control in Brazil (not applicable in South Africa), the UAV equipped with a 10-kg granule tank demonstrates a theoretical field efficiency of 16.5 ha h^{-1}, achieving an operational rate of 3 to 3.5 ha h^{-1} (Programa Cooperativo sobre Mecanização e Automação Florestal, 2021). This reduced performance can be attributed to a low operational efficiency, estimated at approximately 25%. Furthermore, the duration of flights remains inadequate for forest operations, with intervals ranging from 2 to 10 minutes. Nevertheless, contemporary UAV models exhibit superior battery efficiency, facilitating extended flight durations.

3.6.4 Aircraft

The application of glyphosate aerially with aircraft (fixed-wing or helicopter) and as a pre-plant application is seldom practised in South Africa (Rolando et al., 2017). The aircraft used is not dedicated to forestry applications but is predominantly utilised in spraying agricultural farms in local areas. Aerial spraying is more efficient and justifiable when herbicides must be sprayed on large re-establishment areas. There are also environmental impacts (e.g. sensitive areas, aquatic areas and residential areas) and risks (e.g. non-target harm and spray drift) associated with aerial spraying, which limit its practical application in forestry conditions (Payne and Thompson, 1992). Aerial applications can be limited due to high cost, which is dependent on flying time as well as meteorological conditions (i.e. the temperature is below 20°C, humidity not below 40%, maximum wind speed 15 km h^{-1}, no inversion layers at the time of spraying) (Timber Pesticide Working Group, 2023). When identifying the applicable aircraft to use and spraying technique to apply, the terrain, application method, payloads, altitude, accessibility, landing area requirements, and number of personnel required to complete the task (DPE, 2022) are considered. For pre-plant spraying operations, a helicopter with a boom spray can be a cost-effective way to apply herbicide to large and densely infested areas. Due caution must be taken to prevent spray drift and non-target damage (Timber Pesticide Working Group, 2023).

3.7 Planting

Silvicultural re-establishment aims to achieve high survival, stand uniformity and rapid growth at relatively low costs (Pallett, 2005). The quality of seedling material, effective site preparation methods, and suitable planting methods play an essential role in achieving the goal.

A good eucalypt seedling size should vary from 25 to 45 cm, have at least pairs of leaves, the absence of symptoms of nutritional deficiency, and a good leaf water potential, which indicates its hydration level (Figure 3.13). The seedling's stem form varies from 2.5 to 4.5 mm, indicating the survival and growth capacity in the field. Finally, the root plug must have a uniform composition and low density to facilitate handling and development. It should be able to absorb and retain water, even in dry conditions. It must be free of pathogens, invasive plants, insects and toxic substances (Albuquerque et al., 2024).

In plantation forestry, tree planting involves carefully placing a plant into adequately prepared soil (Figure 3.14). The root plug must be entirely in contact with the soil, with a burial of a maximum of 2 cm of the root collar being tolerated. Deep, high, or tilted planting leads to a non-successful planting rate (Albuquerque et al., 2024).

Plant mortality can be caused by poor seedling stock quality, environmental stress (e.g. drought or high temperatures), improper planting handling, and poor planting practices (Jooste, 2022).

In South Africa and Brazil, there are three main methods for planting forests: manual planting tubes, tractor-towed transplanters, and self-propelled and mechanised transplanters (Ramantswana et al., 2020, Sereghetti, 2016). These methods facilitate the successful establishment and growth of seedlings in the ground.

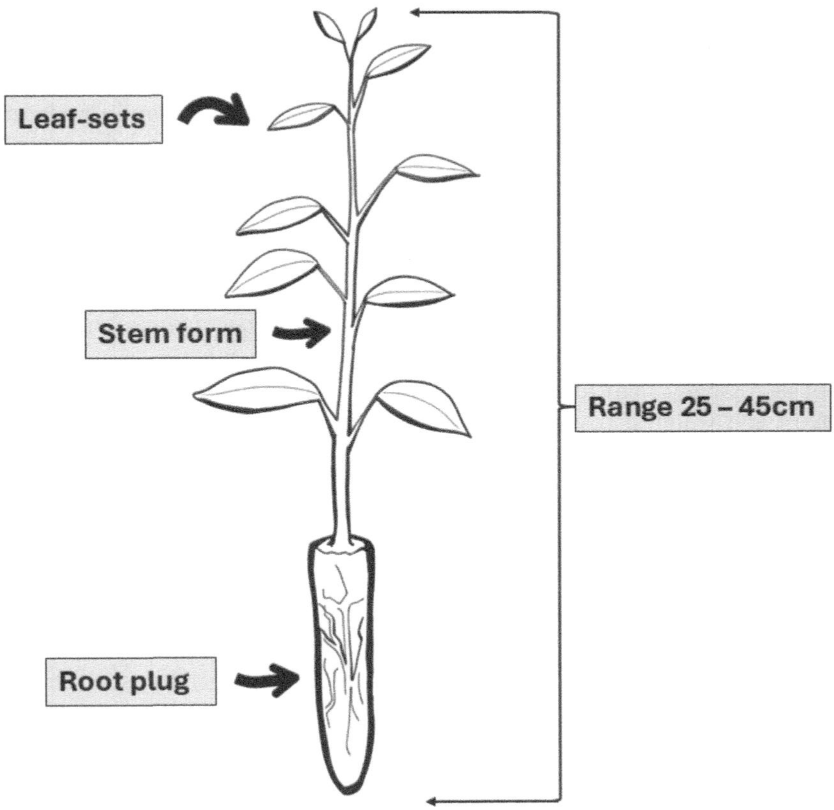

Figure 3.13 General characteristics of a seedling

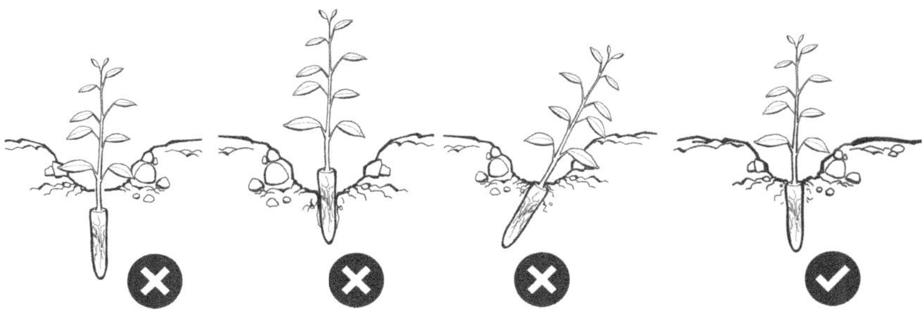

Figure 3.14 Deep, high, tilted and correct planting

Planting can be executed with water, hydrogels and dry planting when conditions allow (Viero and du Toit, 2012). Planting with water and hydrogels aids in the survival and growth of seedlings (Viero and Little, 2006), even though there are reported studies where hydrogels have a negative or zero effect on survival (Crous, 2017). The impact of hydrogel application at planting can be affected by chemicals present in the soil, hydrogel quantity applied, type of polymer, hydrogel particle size distribution, soil texture, physical restrictions to hydrogel expansion, hydrophilic nature of hydrogels, unsaturated hydraulic conductivity between the substrate and the hydrogel particles, application method, planting conditions and rewatering (Crous, 2017). Planting without water can only be practised in areas where the soil is wet or when good rainfall is experienced during the rainy season. In South Africa, the planting windows vary, and they aim to avoid frost in winter and excessive evaporation and heat stress in the summer. The rainfall patterns guide when trees should be planted; however, seedling mortality is reduced through planting seedlings when the soil is moist, planting on cooler days (not excessively high temperatures), and using well-hardened plants which have been watered before in-field dispatch (Evans and Turnbull, 2004). In South Africa, containerised plants are used to re-establish harvested forest stands. The main planting window is from August/September to April/May (applicable to northern KwaZulu-Natal, KwaZulu-Natal midlands, Eastern Cape midlands, and Mpumalanga); the least planting occurs from June to August (applicable to KwaZulu-Natal coastal) (Ramantswana et al., 2020, Viero and du Toit, 2012). Any planting is promptly followed by a blanking operation undertaken between four and eight weeks after planting if the mortality exceeds 5%–10% of the planted crop (Viero, 2007).

3.7.1 Manual planting

In South Africa, manual planting with a trowel or hoe is the most practised method (44.8% of the total 36,753 ha surveyed) (Ramantswana et al., 2020). In Brazil, it is no longer a conventional method of planting; only 0.89% of the 2,796,669 ha surveyed for eucalypt forest plantations in 2022 were manually planted. For pine forest plantations, there is a similar scenario as in South Africa; practically 58% of 223,279 ha surveyed was manually planted (Guerra et al., 2023). When dry planting, the trowel or hoe is used to open the planting hole (depth of 3–5 cm deeper than the plug height) by temporarily displacing the soil from the pit (or rip line) and placing the root plug into the centre of the hole. When practising water planting, materials such as water (1–2 litres in South Africa and up to 4 litres in Brazil) (Wilcken et al., 2011) and hydrogel must be added to the planting position before introducing the seedling (Viero and du Toit, 2012). After that, the displaced soil is carefully deposited back into the hole and firm around the seeding root plug, thereby supporting the plant in remaining upright (Figure 3.15). Containerised clonal plants require careful handling as the root systems are sensitive. When using the Unigro tray with inserts, the planter taps it to loosen and remove the plant. When using the paper pot trays, the plants come in paper sleeves, which keep the root plug intact and reduce the risk of damage. The paper sleeve around the root plug biodegrades gradually after planting (Jooste, 2022).

When planting manually, the following practices are essential:

- Ensure the plant is in good health and meets the plant quality requirements
- Root plugs must be hydrated (Avoid planting plants with dry root plugs)
- Ensure good root-to-soil contact by reducing any clods within the planting pit
- Pack moist soil firmly but gently around the roots to anchor the plant
- Rewatering/irrigation is essential in drought conditions – rewatering can be conducted after 7–14 days, depending on the prevailing conditions. During drought periods with high temperatures, rewatering/irrigation may occur up to 5–7 times after planting in Brazil

Manual planting is conducted in teams comprising a supervisor, manual planter, and additional labour to support the planters. The extra labour is responsible for preparing the seedlings on the roadside and replenishing water/hydrogel and seedlings as and when needed by the planters dispersed around the planting site (Soler et al., 2019). For better quality control, each planter is assigned a specific line to plant, and the team works progressively together to complete planting the area. When water/hydrogel is supplied to the planters, strategic locations are identified to place the water drums used

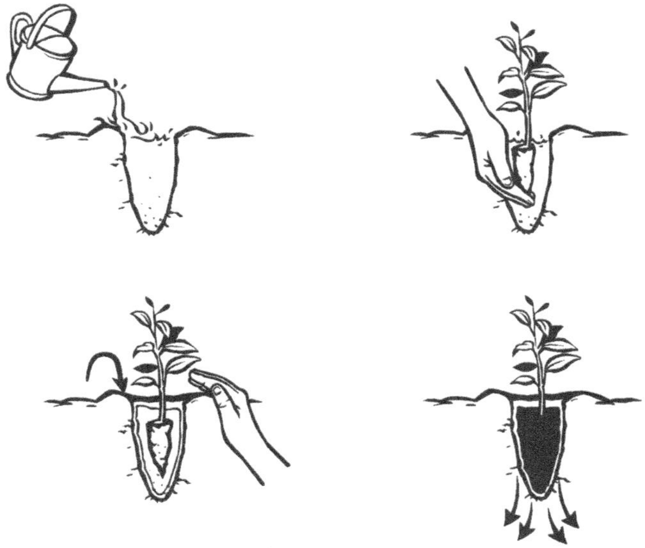

Figure 3.15 Manual water planting process

for mixing the hydrogel. Planning and preparation are critical to avoid operational delays (for example, labour waiting 15–20 minutes for the hydrogel mix to hydrate).

Manual planting productivity may be influenced by factors such as on-site residue conditions (burnt, mulched, broadcasted), terrain (ground roughness and slope), dry or water-based planting, spacing, additional tasks (e.g. adding fertiliser tablets), soil preparation methods (e.g. manual pitting, rip lining), team size, working techniques, and procedures for replenishment (water and seedlings). Productivity per tree can range from 0.7 minutes (85 trees ha^{-1}) to 1.3 minutes (46 trees ha^{-1}), depending on these factors. In Brazil, Lemos (2014) evaluated forestry workers who planted eucalyptus in previously subsoiled areas, achieving a yield of 357 seedlings per hour with a 3 × 1.5 m spacing.

Manual planting produces better quality than mechanised planting (Nieuwenhuis and Egan, 2002). The seedling planting process can be affected by several micro-level factors, which an individual can assess and take remedial actions for. For example, if the pit contains too many clods, a person can break them up and reduce air pockets, thereby enhancing soil and root contact (Figure 3.16). A planting machine may not be able to accommodate the small site variability common in the area.

3.7.2 Planting with planting tubes

In South Africa, planting tubes are the second most practised method (44.8% of the total 36,753 ha surveyed) (Ramantswana et al., 2020). In Brazil, planting with tubes remains the most popular method for establishing

Figure 3.16 Manual planting

eucalyptus forest plantations. In 2022, it represented almost 62% of the 2,796,669 ha surveyed eucalypt plantations. For pine forest plantations, there is a similar scenario as in South Africa; practically 42.3% of 223,279 ha surveyed was planted with tubes, being the second most practised method (Guerra et al., 2023). There are different types of planting tubes available. A planting tube is about 1 m long and varies in width depending on the seedling root plugs and species form. Planting tube configurations vary based on whether the planter has irrigation or non-irrigation functionalities.

Types of planting tubes

- Plant only – this type of planting tube is used for planting only; it is not designed to facilitate the application of water or hydrogel at planting
- Planting tube and independent irrigation (disintegrated) – this planting tube is designed to facilitate planting with the support of another person, adding water or hydrogel into the planting tube beak, e.g. Krupps' planter, Vastrap
- Backpack with planting tube (integrated unit) – this planting tube is designed to facilitate planting and irrigation from the same unit. The individual carrying the backpack with water can activate the irrigation during the planting process, e.g. Fakka Jelly. The planter may also carry a seedling bucket
- Planting tubes integrated into a carrier vehicle or tank – several planting tubes are connected to a boom mounted on the carrier vehicle or tank. Details of this configuration are provided in the section that describes the semi-mechanised planted method

When using planting tubes, the team comprises a supervisor, planter(s) (person who operates the planting tube), planter assistants and additional people responsible for supporting the planters with replenishment of water (or hydrogel) and seedlings (including preparation on the roadside). The planter typically carries seedlings in a bucket; however, in some cases, the planter works with an assistant responsible for irrigation (water or hydrogel) and seedling supply. When dry planting, the beak of the tube is driven into the soil by the planter. The beak blades are opened by pressing on the pedal with the foot, followed by dropping the seedling through the tube before the unit is lifted, and the soil around the plant is firm. When planting with an irrigation-compatible unit, the operator inserts the beak of the tool into the pit and opens the beak only marginally, then inserts the seedling into the planting tube (Figure 3.17). The planter pulls the trigger to allow the water flow (or hydrogel). Once the water (hydrogel) fills the beak and flows over the top, the required water/gel (estimated litres) will have been dispersed into the pit with the seedling. As the beak is pulled out of the soil, it must be opened fully to allow the seedling to be left behind. After that, the planter

Figure 3.17 Manual planting with a planting tube with irrigation

lets go of the left arm, and the tool closes automatically. Before proceeding, the planter covers the root plug with soil and lightly taps the surface to firm seedlings and remove any air pockets.

When using planting tubes without irrigation, planting productivity varies from just over 300 seedlings per hour under challenging conditions to just under 500 trees per hour in relatively easy conditions (Appelroth, 1971). In Brazil, a study conducted by Fessel (2003) evaluated the manual planting operation, involving 13 people, using planting tubes without irrigation, and achieved 1,240 seedlings per hour. When using planting tubes integrated with irrigation, productivity can range from 200 to 400 trees per hour (Ramantswana, 2023). High productivity of up to 1,600 trees per hour can be achieved when working in flat, sandy areas with no residues by using an adapted pitless planting tube system (referred to as the Vastrap), which comprises a manual planter with an assistant responsible for irrigation only (Khoza, 2023) (Figure 3.18). Planting tubes improve ergonomics compared to conventional manual planting because the worker does not have to bend down or bend knees when planting (Appelroth, 1971). However, workers report the most significant discomfort during planting, mainly when using manual planters attached to the backpack containing hydrogel, which can weigh nearly 30 kg, and carrying a box of seedlings (Soler et al., 2019). Essentially, the goal is to reduce extreme postures associated with the planting task

Figure 3.18 Pitless planting system

(Parker, 2013), so mechanisation is a potential solution to these exhausting physical and mental operations for workers (Souza, 2002).

3.7.3 Semi-mechanised planting

Semi-mechanised planting refers to a combination of manual and machine planting. The planting is undertaken manually using a planting tube system combined with a machine that efficiently carries resources (e.g. water, seedlings) required for planting. In South Africa, semi-mechanised planting is the third most practised method (16.1% of the total 36,753 ha surveyed) (Ramantswana et al., 2020).

Different types of semi-mechanised planting systems can be differentiated by the carrier's role, planter configuration, and terrain application.

- A tractor/dozer-drawn planting trailer is pulled and guided from a single hitch point. The drawing machine does not entirely support the drawn trailer
- Tractor/dozer-mounted planting – the planting implement is attached to the tractor with a hitch linkage so that it is entirely supported by the tractor
- Self-propelled planting comprises a single machine with its source of tractive power for propulsion (Figure 3.19)

A semi-mechanised planting system comprises four to seven employees walking behind a water trailer carrying a planting tube and seedlings. When

Figure 3.19 Self-propelled semi-mechanised planting system

planting, the planters work in a line behind the drawing/carrier machine, which travels slowly ahead of the planters. Each planter walks to a planting location, inserts the planting tube into the prepared soil, deposits the seedlings into the planting tube, and then activates the irrigation before slowly removing the planting tube, firming, and moving to the next planting spot. The irrigation is supplied from a tank integrated into the planter's planting tubes. The planters carry seedlings in a bucket regularly replenished by a person responsible for providing the seedlings. When using semi-mechanised planting systems (planting only) with integrated irrigation, the planting productivity varies between 8 ha and 11 ha per shift (9-hour shift). Machine configurations, spacing, planting line distance, road density, and residues can influence productivity.

Semi-mechanised planting can also integrate additional tasks, such as soil preparation before planting (e.g. Wasserplanter). High-pressure water can create a planting spot before the seedlings are planted.

The advantages of using semi-mechanised planting:

- More productive than manual methods
- It is more ergonomically friendly
- Efficiency in the supply of resources (e.g. water and seedlings) required for planting, as the machine carries them
- Less complex supervision as the operation is consolidated, and work is undertaken progressively as a unit
- Easier management of containers/trays

Disadvantages of using mechanised planting:

- Terrain limitations if using non-purpose-built equipment
- More capital-intensive compared to manual methods

- Planting quality may be influenced negatively, as planters may not dedicate time to planting properly due to the pressure of other planters
- Limited flexibility – if the carrier machine breaks, then the whole planting operation stops

3.7.4 Mechanised planting

The history of mechanised planters in Brazil is recent, representing a shift from mostly manual planting methods to advanced automated machines and systems. Tree planting remains a manual task, where labourers depend on simple tools such as the *chucho* (a pointed metal stick) or a manual transplanter called the *matraca* (a metal tube with a manual opening at the bottom) to place the seedling into the ground. So far, an average worker can plant around 1,200–1,400 plants per day (an 8-hour shift), depending on the soil type, slope, and seedling quality.

In the 1990s, there were some tractor-pulled planters, where one worker sat on the planting machine (while another operated the tractor) and fed seedlings into the planting tube. These machines were sometimes capable of ripping the soil, planting, adding fertiliser, and spraying herbicides, or any combination of these operations as designated by the forest industry. Those machines were locally manufactured and maintained, supported by national suppliers, and could cover 2–4 hectares daily, working an 8-hour shift. However, from the 1990s to the 2000s, there was no disruptive innovation in silviculture operations nor a shortage of workers. However, as the new century began, labour costs rose, demand for timber increased, and new forest plantations started in the western part of the country.

Back in 2010, the first fully mechanised planter was imported from Sweden by the Institute of Forest Research and Studies (IPEF) under the Cooperative Programme on Forest Mechanisation and Automation (PCMAF). It was attached to a hydraulic excavator for the first time in South America. This seedling planter could simultaneously plant, fertilise, and irrigate, although soil preparation was done previously by another equipment. Initially, some planting quality issues were observed, such as poorly fixed seedlings and covered root collars. The field trials showed average productivity from 320 to 350 seedlings PMH^{-1}, being more expensive than manual or semi-mechanised alternatives. It´s important to note that seedling quality and soil conditions highly influenced machinery performance for the first time.

After several years of collaborative development and testing of new solutions, the Brazilian forestry sector, universities, and manufacturers often matched manual planting in overall quality but still struggled with spacing deviations. During the following years, from 2010 to 2025, several innovations emerged in Brazil's forestry sector. One was the development of an automated triple planter, the first of its kind in the world, which used a tracked dozer as its base and was equipped with three unique planting heads, capable of planting and irrigating three seedlings per cycle. One primary concern was introducing a high-precision georeferencing system based on RTX (Real Time Extended), accurate to within 5 cm. The triple automated planter productivity ranges from 650 to 800 seedlings per hour (including refuelling) and up to 900 seedlings per hour without refuelling (Figure 3.20). For the first time, fully mechanised systems surpassed all previously documented planting field performance, with spacing deviations of less than 1% from the target. One drawback of this latest planting system is the difficulty in working on uneven terrains. Therefore, heavier excavators (24 tonnes) were chosen to be attached to a new single planting header, allowing a vertical positioning system to perform deep subsoiling and plant the seedling in a vertical position on steep terrain, add fertiliser and water through the planter foot.

Recently, another tree planting machine was introduced in Brazil and South America, contributing to the broader development of mechanised planting (Figure 3.21). A forwarder-based carrier of a tree planting system combines soil preparation, seedling planting, fertilisation, irrigation, and georeferencing solutions with two planting arms positioned at the rear of the machine. This model is designed for large-scale reforestation and continuous

Figure 3.20 Triple automated planter

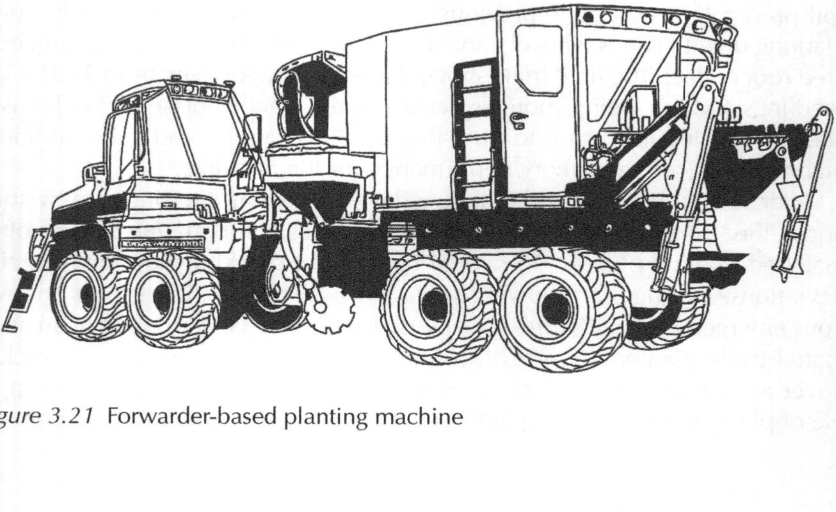

Figure 3.21 Forwarder-based planting machine

planting cycles, minimising stops for seedling reloading, and represents a disruptive innovation, suitable for flat to moderately uneven terrain. It has high productivity, achieving an average of 2,000 seedlings PMH^{-1} under ideal soil conditions, the maximum rate recorded to date. Brazil is actively engaged in assessing and adopting innovative planting technologies from around the world, as well as from local manufacturers.

South Africa lacks mechanised planting operation machine options. In 2018, an excavator-based planting machine called the Proplant 1 (PP1) was tested in forestry conditions; however, this version of the machine was not commercialised due to reliability issues. The machine can pit, apply a pre-emergent spot spray, plant one seedling, add water or gel to the pit, add fertiliser, and firm (Novelquipforestry, 2024). In 2025, an excavator-based planting machine with an imported single planting head attached was introduced and is currently undergoing testing. The concept machine of the PP1 has recently (May 2025) been reintroduced to the local South African market, and it is undergoing testing. The challenges with introducing capital-intensive planting equipment in South Africa include a limited overall planting forest area, small compartments/stands, a limited planting season confined to specific periods, and the availability of labour in certain areas (lack of employment).

In Eswatini (a landlocked country bordered by South Africa in its north, south, west, and southeast boundaries), a fully mechanised excavator-based single-head planting machine was introduced in 2022 (Figure 3.22). The machine can pit (based on the GPS marking system), plant, and irrigate one plant at a time. The productivity of the planting machine is estimated to vary between 150 and 250 plants per hour, depending on residue conditions, stump heights, the length of the planting lines, and the operator's experience (Zwane, 2023). The machine's application has faced challenges, including low production, suitability of terrain conditions, frequent relocation, lack of

Figure 3.22 Excavator-based single-head planting machine

local expertise in machine maintenance, a short planting window, and complexity in planting at night (Zwane, 2023). Very little scientific information is available on the plants' survival, growth, and uniformity established by the planting machine.

3.7.5 Mechanised irrigation

In Brazil, irrigation has become a crucial operation in establishing eucalyptus plantations, serving as a limiting factor for seedling survival rates, especially during dry seasons or in the north, northeast, or midwest regions. Over the last 20–25 years, semi-mechanised systems have been widely used, and many local manufacturers or metal shops have been assembling irrigation systems for the forest sector. In field operations, it is commonly carried out using a tractor-drawn water tanker (3,000–7,000 litres), connected to the labourers via multiple hoses (3–7). A typical operation involves one tractor driver (e.g. a truck or a forwarder base machine) and three, five, or seven labourers, conducting two to four irrigation cycles after planting. They deliver 3–4 litres of water per seedling, initially right after planting and then weekly for up to 30–45 days.

Nowadays, mechanised irrigation systems use tractor-mounted implements with one, two, or three nozzles, allowing simultaneous irrigation. Mechanical systems aim to reduce labour and improve irrigation uniformity, thereby decreasing runoff. However, newer solutions with artificial intelligence, computational vision, and embedded computers onboard introduce high operational costs, still experiencing unproductive time due to frequent refilling of water tanks (small capacities) and mechanical downtime caused by agricultural tractors not being designed for silvicultural operations. To reduce costs, reduce irrigation frequency and improve water use efficiency, adding hydrogel can extend soil moisture retention. Some polymers can extend the irrigation interval from 24 hours to up to 10 days, depending on the weather scenario, thereby improving seedling establishment. It is also positive for the environment, significantly reducing the initial irrigation volume from 3–4 litres to 0.25–0.5 litres per seedling, optimising the water–hydrogel tank refilling frequency. Adopting irrigation with hydrogel potentially improves labour efficiency by up to 40%, allowing a worker to irrigate 1.3 hectares per day, instead of 0.7 hectares per day.

3.8 Fertilising

Plant nutrition is an essential activity in intensive plantation forestry. The aim of fertilising in plantation forestry is to provide nutrients needed for the plant to grow. It is considered the simplest way to improve soil fertility, correct specific nutrient efficiency, correct poor soils, and stimulate growth. Plants require both macronutrients (e.g. nitrogen (N), phosphorus (P), potassium (K), calcium (Ca), magnesium (Mg) and sulphur (S) and micronutrients (e.g. iron

(Fe), copper (Cu), chlorine (Cl), boron (B) manganese (Mn), Zinc (Zn), molybdenum (Mo)) (Evans and Turnbull, 2004). When these nutrients are lacking, plant growth may be negatively affected. When fertilising plants in South Africa, especially eucalypts, various factors such as soil organic matter, soil cultivation method, slash management, and weed control can influence the nutrient requirements on a particular site (Maree et al., 2012). The forest stand is composed of large nutrient pools, which can potentially be affected by various forest management practices (Smith, 2012).

3.8.1 Manual fertilising

Fertilisation improves early growth in eucalyptus, compensating for the nutritional limitation of the soil. However, applying fertiliser requires an investment, and fertiliser accounts for a high percentage of the total cost of establishing a new forest (Soler et al., 2019). Over 60% of the re-established land in South Africa is fertilised manually (Figure 3.23). Fertilisers can also be applied using fertiliser forks (Figure 3.24) and backpacks (Ramantswana et al., 2020). Manual fertilising is Brazil's second most popular method for eucalypt forest plantations. In 2024, it represented almost 27% of the 1,575,381 ha surveyed eucalypt plantations (Guerra et al., 2025). For pine forest plantations, 30% of 223,279 ha surveyed was manual fertilising (Guerra et al., 2023). Fertiliser is applied in either granular or tablet form. The tablets are commonly placed 15 cm away from the transplant, and granular fertiliser is placed in two slots 15 cm away (Figure 3.25). Manual fertiliser application is slow, but it guarantees that the fertiliser is placed in the correct location, next to the tree. Fertiliser must not be placed too close to the seedling, burning the roots or too far from where it is inaccessible to the plant (Evans and Turnbull, 2004).

Figure 3.23 Manual fertilising

Figure 3.24 Manual fertilising with a fertiliser fork

Figure 3.25 Manual fertilising using fertiliser tablets

3.8.2 Mechanised fertilising

Mechanised fertilising is Brazil's most popular method for eucalypt forest plantations. Since adopting mechanised planters in Brazil, some manufacturers have developed solutions for fertilising seedlings during the planting operation, adding 100–150 g per seedling of 06–30–06 NPK (Wilcken et al.,

2011). In 2024, it represented almost 67% of the 1.575.381 ha surveyed. It is interesting to note the 1% of automatic fertilisation on the same total area surveyed (Guerra et al., 2025). For pine forest plantations, 70% of the 223.279 ha surveyed were mechanically fertilised, according to Guerra et al. (2023).

Mechanised fertilising is a standard practice in eucalypt forest plantations in Brazil, aimed at optimising tree growth and productivity (Castro, 2017); it is usually carried out using an adapted agricultural tractor coupled with a fertiliser spreader (Figure 3.26). Sometimes, the fertiliser spreader incorporates embedded technology to control the amount of fertiliser released precisely and supports creating a map of the worked area according to precision forestry principles. On other occasions, the fertiliser spreader applies a continuous or intermittent strip along the ripped or planting line (Wilcken et al., 2011). There are various types of fertilisation, including planting fertilisation

(a)

(b)

Figure 3.26 Mechanised fertiliser equipment

(e.g. mechanised planting and fertilising heads, tractors coupled with a fertiliser spreader, forwarders carrying a spreader); top-dressing fertilisation and maintenance fertilisation (e.g. tractors with a fertiliser spreader, forwarders carrying a spreader, or even drones).

In areas with minimum tillage, such as flat or gently undulating soil, a multifunctional subsoiler is commonly used. This implement can perform multiple tasks simultaneously, including cutting residues and tree roots, subsoiling, and applying base fertiliser along with pre- and post-emergent herbicides (Castro, 2017). Combining functions such as subsoiling and fertilising, or fertilising and herbicide application, within a single piece of equipment is considered a cost-effective way to mechanise regeneration activities (Soler et al., 2024).

3.9 Regeneration through coppice

Coppice regeneration is a method used to re-establish eucalypts after clear-felling. Coppice regeneration is a more cost-effective and faster way to re-establish a eucalyptus stand because it saves on the conventional costs of residue management, soil preparation, and planting a new crop. For optimal coppicing yield, the correct species with vigorous growth, proper spacing, good survival of the original stand, and limited harvesting damage on stumps must be considered (Crous and Burger, 2015). The coppice shoots can resprout from the stump, and the best shoots can grow through selection. The process of reducing the coppice to leave the desired ones is performed in two stages. In the first stage, the shoots are reduced to two per stump when the dominant height is between 2 and 4 m. After that, the final reduction to one shoot per stump is conducted when the dominant height reaches 8 m (Viero and du Toit, 2012).

The coppice reduction can be conducted manually or chemically. Manual coppice reduction can be performed using a hatchet or bush knife to cut the unwanted shoots; if not practised carefully, the selected shoots can be damaged (Little, 2007, Roberts et al., 2017). The chemical treatment of coppice reduction involves the application of glyphosate (a systemic herbicide) on the leaves of unwanted coppice, while being cautious not to spray the selected coppice stem (Little and Toit, 2003). This method can be more cost-effective than manual methods in controlling secondary coppice regrowth (Little, 2007, Roberts et al., 2016). Motor manual clearing saws can also cut shoots during the coppice reduction process.

3.10 Summary

This chapter explored silviculture technologies used in establishment and re-establishment. It detailed residue management techniques, such as controlled burning, mulching, and windrowing, and noted the impact of harvesting methods on residue levels. The document examined methods for marking and preparing planting positions, including manual and mechanised pitting and subsoiling. It covered weed control strategies applied before planting, including manual and tractor boom chemical sprays, as well as the use of UAVs. Various planting technologies were discussed, from manual planting to semi-mechanised and mechanised approaches, including automated systems. This chapter also addressed fertilisation methods, differentiating between manual and mechanised techniques, and noted the use of hydrogels for retaining irrigation moisture. Finally, it touched on regeneration through coppice, including methods for coppice reduction. Overall, this chapter outlined the importance of matching technologies to specific site and stand requirements to achieve successful forest regeneration.

Review questions

1 Choose an example of a mechanised residue management technology and describe its application, advantages, and disadvantages
2 Choose an example of a mechanised preparation of a planting position technology and describe its application, advantages, and disadvantages
3 Choose an example of a mechanised weeding before planting technology and describe its application, advantages, and disadvantages
4 Describe the various available planting technologies by focusing on differences in configuration and applications
5 Explain the advantages and disadvantages of mechanised fertiliser application in regeneration operations

References

Achat, D. L., Deleuze, C., Landmann, G., Pousse, N., Ranger, J. & Augusto, L. 2015. Quantifying consequences of removing harvesting residues on forest soils and tree growth – A meta-analysis. *Forest Ecology and Management,* 348, 124–141.

Ackerman, S., Fryer, D. & Jackson, R. 2018. *South African Outdoor Power Equipment Safety & Operating Handbook*. Pietermaritzburg: Institute for Commercial Forestry Research.

Albuquerque, A. C., Oguri, G., Balista, B. Z., Ramantswana, M. & Guerra, S. P. S. 2024. *Overview of Eucalypt Plantation in Brazil*. Botucatu: FEPAF.

Alcázar, J., Woodard, P. M. & Rothwell, R. L. 2002. Soil disturbance and the potential for erosion after mechanical site preparation. *Northern Journal of Applied Forestry,* 19, 5–13.

Allan, R., Morris, A. & Carlson, C. 2000. Survival and early growth effects of some re-establishment pratices with Pinus patula. *Southern African Forestry Journal,* 187, 29–36.

Alwi, A., Lapammu, M., Jarapudin, Y., Molony, K., Boden, D., Macdonell, P., Warburton, P., Brawner, J. & Meder, R. 2020. Importance of weed control prior to planting for the establishment of planted forests in Sabah, Malaysia. *Journal of Tropical Forest Science,* 32, 349–354.

Amaral, L. R. D., Zerbato, C., Freitas, R. G. D., Barbosa Júnior, M. R. & Simões, I. O. P. D. S. 2021. UAV applications in Agriculture 4.0. *Revista Ciência Agronômica,* 51, e20207748.

Appelroth, S. 1971. Planting tube makes it easy to plant Japanese paperpot planting stock in Finland. *The Forestry Chronicle,* 47, 350–351.

Arjomandi, M., Agostino, S., Mammone, M., Nelson, M. & Zhou, T. 2006. Classification of unmanned aerial vehicles. *Report for Mechanical Engineering Class*, University of Adelaide, Adelaide, Australia, 1–48.

Baptista, J. 2008. *Métodos de preparo de solo e sua influência na erosão hídrica e desenvolvimento inicial de Eucalyptus saligna em um cambissolo háplico da depressão central do estado do rio grande do sul.* Masters dissertation in Soil science, Universidad Federal de Río Grande del Sur.

Berry, M. & Sessions, J. 2020. Evaluating the economic incentives of biomass removal on site preparation for different harvesting systems in Australia. *Forests,* 11, 1370.

Bigley, R. & Henderson, J. 1989. Effects of slash burning on plant succession and timber stand establishment. *In*: Hanley, D.P., et al. (eds.) *The Burning Decision: Regional Perspectives on Slash*. Seattle: College of Forestry Resource, University of Washington, 47–68.

Bone, E. & Bolkcom, C. 2003. *Unmanned Aerial Vehicles: Background and Issues for Congress*. SE Washington: Foreign Affairs, Defence and Trade Division.

Bugatti Ig, M. D., Martins, D. S., & Marconato, E. S. 2012. Automação e controle de pulverização em máquinas agrícolas. *e-F@tec*, 2 (1), 1–10.

Carvalho, F. K., Chechetto, R. G., Mota, A. A. & Antuniassi, U. R. 2020. Challenges of aircraft and drone spray applications. *Outlooks on Pest Management,* 31, 83–88.

Castro, D. C. V. 2017. *Planejamento da qualidade dos processos silviculturais na cultura do eucalipto no Brasil.* Sãu Paulo: Universidade de Sãu Paulo.

Chapman, C. 2012. Mechanised pitting machine in the spotlight. *SA Forestry.* Hillcrest: Green Forest Media.

Chapman, C. 2020. Mulching operations bring benefits for growers. *SA Forestry.* 3 Minerva Rd: Green Forest Media (Pty) Ltd.

Chapman, C. 2021. *Pioneering Aerial Drone Spraying in Forestry* [Online]. Hillcrest: Green Forest Media (Pty) Ltd. Available: https://saforestryonline.co.za/articles/pioneering-aerial-drone-spraying-in-forestry/#:~:text=An%20innovative%20silviculture%20contractor%20in,is%20practiced%20in%20South%20Africa. [Accessed 26 March 2024].

Chudakov, D. & Okúlik, D. 1977. *Fundamentos de la teoría y el cálculo de tractores y automóviles*. Moscú: Editorial MIR.

Clarke, J., Kelly-Quinn, M., Blacklocke, S. & Bruen, M. 2015. The effect of forest windrowing on physico-chemical water quality in Ireland. *Science of the Total Environment,* 514, 155–169.

Crane, J. 2017. A decade of developing novel mechanised silviculture equipment. *Wood Southern Africa and Timber Times.* Johannesburg: Malnor Publishing.

Crous, J. W. 2017. Use of hydrogels in the planting of industrial wood plantations. *Southern Forests: A Journal of Forest Science,* 79, 197–213.

Crous, J. W. & Burger, L. 2015. A comparison of planting and coppice regeneration of Eucalyptus grandis× Eucalyptus urophylla clones in South Africa. *Southern Forests: A Journal of Forest Science,* 77, 277–285.

Damilare. 2023. *Motorized Knapsack Sprayer – A Detailed Guide for Efficient Use* [Online]. Online: Transvilleagrong. Available: https://transvilleagrong.com/motorized-knapsack-sprayer-a-detailed-guide-for-efficient-use/ [Accessed 22 March 2024].

De Melo, R. O., DA Fonseca, A. A., De Barros, N. F., Fernandes, R. B. A., Teixeira, R. D. S., Melo, I. N. & Martins, R. P. 2022. Retention of eucalyptus harvest residues reduces soil compaction caused by deep subsoiling. *Journal of Forestry Research,* 33, 643–651.

de Moraes Gonçalves, J. L., Stape, J. L., Laclau, J.-P., Smethurst, P. & Gava, J. L. 2004. Silvicultural effects on the productivity and wood quality of eucalypt plantations. *Forest Ecology and Management,* 193, 45–61.

Del Cerro, J., Cruz Ulloa, C., Barrientos, A. & De León Rivas, J. 2021. Unmanned aerial vehicles in agriculture: A survey. *Agronomy,* 11, 203.

DPE. 2022. *Aerial Spraying Guidelines*. New South Wales: Environment and Heritage.

Du Toit, B., Smith, C. W., Little, K. M., Boreham, G. & Pallett, R. N. 2010. Intensive, site-specific silviculture: Manipulating resource availability at establishment for improved stand productivity. A review of South African research. *Forest Ecology and Management,* 259, 1836–1845.

Espinoza, J. A. 2004. Site selection, site preparation, and weed control for Gmelina arborea in western Venezuela. *New Forests,* 28, 217–226.

Evans, J. & Turnbull, J. W. 2004. *Plantation Forestry in the Tropics: The Role, Silviculture, and Use of Planted Forests for Industrial, Social, Environmental, and Agroforestry Purposes*. Oxford: Oxford University Press.

Fessel, V. A. G. 2003. *Qualidade, desempenho operacional e custos de plantios, manual e mecanizado, de Eucalyptus grandis, implantados com cultivo mínimo do solo.* Piracicaba, São Paulo: Universidade de São Paulo.

Gatto, A., Barros, N. F. D., Novais, R. F. D., Costa, L. M. D. & Neves, J. C. L. 2003. Efeito do método de preparo do solo, em área de reforma, nas suas características, na composição mineral e na produtividade de plantações de Eucalyptus grandis. *Revista Árvore,* 27, 635–646.

Ghaffariyan, M. R. & Dupuis, E. 2021. Analysing the impact of harvesting methods on the quantity of harvesting residues: An Australian case study. *Forests,* 12, 1212.

Gonçalvez, J. L. M. 2002. Conservação do solo. *In*: Gonçalves, J. L. M., S. J. (ed.) *Conservação e cultivo de solos para plantações florestais*. Piracicaba: IPEF, 47–129.

Greyling, J. 2022. A crop spraying drones an option? *Koringfokus.* Noordbrug: Mediacom.

Guanfeng. 2022. *What Are the Types of Knapsack Sprayers?* [Online]. Zhejiang: Taizhou Gaungfeng Plastic. Available: https://www.gfsprayer.com/What-are-the-types-of-knapsack-sprayers-id45856477.html [Accessed 21 March 2024].

Guerra, S. P. S., Oguri, G., Schumacher, M. V., Kulmann, M. S. D. S. & Faria, J. I. 2023. *Survey of the Level of Mechanization in Forestry: 2022/2023 Edition*. Piracicaba: IPEF.

Guerra, S. P. S., Oguri, G., Schumacher, M. V., Witschoreck, R. & Balista, B. Z. 2025. *Survey of the Level of Mechanization in Forestry: 2024/2025 Edition*. Piracicaba: IPEF.

Halbrook, J., Han, H., Graham, R., Jain, T. & Denner, R. 2006. Mastification: A fuel reduction and site preparation alternative. *In:* Chung, W. & Han, H. (eds.) *The 29th Council on Forest Engineering Conference.* Idaho: COFE.

Hechter, U., Little, K. M. & Titshall, L. 2020. The influence of manual and motor-manual pitting implements, pit size and quality on eucalypt performance, South Africa. *Southern Forests: A Journal of Forest Science,* 82, 170–178.

IBGE. 2023. *PEVS – Produção da Extração Vegetal e da Silvicultura* [Online]. Available: https://www.ibge.gov.br/estatisticas/economicas/agricultura-e-pecuaria/9105-producao-da-extracao-vegetal-e-da-silvicultura.html [Accessed 16 February 2024].

Jacob, Z. 2022. *The effects of slash management methods and pit preparation techniques on pit quality, growth response, operational productivity, and its cost effectiveness at re-establishment.* Master of Science Forestry Dissertation, Nelson Mandela University.

Jones, B. J. 2006. *Mechanised Spraying Systems for Herbicide Use in Forestry*. Edinburgh: Forestry Commission.

Jooste, A. (ed.) 2022. *Tree Farming Guidelines*. Cascades: Sappi.

Kanowski, P. 1997. *Afforestation and Plantation Forestry*. Canberra: Resource Management in Asia-Pacific.

Keefe, R. F., Zimbelman, E. G. & Picchi, G. 2022. Use of individual tree and product level data to improve operational forestry. *Current Forestry Reports,* 8, 148–165.

Khoza, M. 2023. *Comparison of the Operational Productivity and Quality of Manual and Semi-Mechanised Planting Techniques KwaZulu-Natal, South Africa*. Port Elizabeth: Honors in Natural Resource Management, Nelson Mandela University.

Kim, J., Kim, S., Ju, C. & Son, H. I. 2019. Unmanned aerial vehicles in agriculture: A review of perspective of platform, control, and applications. *IEEE Access,* 7, 105100–105115.

Lemos, S. V. 2014. *Análise econômica do plantio e condução do eucalipto de curta rotação para fins energéticos*. Dissertação de Mestrado, Universidade Federal de Viçosa, Viçosa, MG, Brasil.

Light, M. 2015. *Plantation Re-Establishment: Principles of Pit Preparation*. Pietermaritzburg: Institute for Commercial Forestry Research (ICFR).

Lima, J. S. D. S., Souza, A. P. D., Machado, C. C. & Pezzopane, J. E. M. 2004. Estimativa das estabilidades longitudinal e transversal de tratores florestais utilizados na colheita de madeira. *Revista Árvore,* 28, 839–844.

Little, K. M. 2007. Final results from a Eucalyptus grandis× E. camaldulensis coppice trial. *Scientia Forestalis,* 76, 97–108. Available: https://www.ipef.br/publicacoes/scientia/nr76/cap08.pdf.

Little, K. M. & Toit, B. D. 2003. Management of Eucalyptus grandis coppice regeneration of seedling parent stock in Zululand, South Africa. *Australian Forestry,* 66, 108–112.

Lopes, E. D. S., Oliveira, F. M. D. & Rodrigues, C. K. 2011. Determinação da carga física de trabalho na atividade de aplicação manual de herbicida Determination of demanded physical effort in herbicide application activity. *Ambiência,* 7, 329–337.

Lyle, G. 1981. Chemical weed control in forestry. *South African Forestry Journal,* 117, 61–64.

Mahadeo, L. 2022. *Understanding the system and supply chain effects of two residue management treatments in pine and eucalyptus plantations*. Masters thesis, Stellenbosch University.

Maree, B., Kotze, H. & du Toit, B. 2012. Silviculture of industrial Eucalyptus and Acacia plantations in South Africa. *In:* Bredenkamp, B. & Upfold, S. (eds.) *South African Forestry Handboook*. 5th ed. Pretoria: South African Institute of Forestry, 63–78.

Mathelele, O. 2019. *Technical survey of the South African harvesting machine population*. Masters Thesis, Nelson Mandela University.

Medauar, C. C., Silva, S. D. A., Carvalho, L. C. C., Tibúrcio, R. A. S., Lima, J. S. D. S. & Medauar, P. A. S. 2018. Monitoring of eucalyptus sprouts control using digital images obtained by unmanned aerial vehicle. *Journal of Sustainable Forestry,* 37, 739–752.

Mialhe, L. G. 1980. Máquinas motoras na agricultura. *EDUSP,* 2, 367.

Morris, L. A., Pritchett, W. L. & Swindel, B. F. 1983. Displacement of nutrients into windrows during site preparation of a flatwood forest. *Soil Science Society of America Journal,* 47, 591–594.

Nadel, R., Scholes, M. & Byrne, M. 2007. Slash burning, faunal composition, and nutrient dynamics in a Eucalyptus grandis plantation in South Africa. *Canadian Journal of Forest Research,* 37, 226–235.

Ndlovu, N., Little, K., Titshall, L. & Rolando, C. 2019. Site preparation and vegetation management impacts on Pinus patula growth and rotation end productivity in South Africa. *Australian Forestry,* 82, 107–115.

Nieuwenhuis, M. & Egan, D. 2002. An evaluation and comparison of mechanised and manual tree planting on afforestation and reforestation sites in Ireland. *International Journal of Forest Engineering,* 13, 11–23.

Novelquipforestry. 2024. *Products* [Online]. George: Novelquip. Available: https://www.nqfsa.com/products/ [Accessed 24 April 2024].

Nsele, M. 2023. Advanced weeding with the Troner tracctor sprayer. *In:* Ramantswana, M., ed. *Modern Silviculture Webinar – Optimising Modern Silviculture,* George (Online). George: Nelson Mandela University.

Nyland, R. D. 2016. *Silviculture: Concepts and Applications.* New York: State University of New York: College of Environmental Science and Forestry, USA, Waveland Press, Inc.

Pacsys. 2024. DJI AGRAS T40 Crop spraying drone info sheet. *In:* PACSYS (ed.) *PACSys sales.* KwaZulu-Natal: PACSys.

Paes, F. A. S. V., Lima, A. M. N., Hakamada, R. E. & Barros, N. F. D. 2013. Impacto do manejo dos resíduos da colheita, do preparo do solo e da adubação na produtividade de eucalipto. *Revista Brasileira de Ciência do Solo,* 37, 1081–1090.

Page-Dumroese, D. S., Busse, M. D., Archuleta, J. G., Mcavoy, D. & Roussel, E. 2017. Methods to reduce forest residue volume after timber harvesting and produce black carbon. *Scientifica,* 2017, 1–8.

Pallett, R. N. 2005. Precision forestry for pulpwood re-establishment silviculture. *Southern African Forestry Journal,* 203, 33–40.

Parker, R. J. 2013. *A field investigation into the impact of task demands on worker responses in the South African forestry silviculture sector.* Master Thesis, Rhodes University.

Payne, N. J. & Thompson, D. G. 1992. Off-target glyphosate deposits from aerial silvicultural applications under various meteorological conditions. *Pesticide Science,* 34, 53–59.

Pereira, D. P., Fiedler, N. C., Lima, J. S. D. S., Guimarães, P. P., Môra, R. & Carmo, F. C. D. A. D. 2012. Eficiência da subsolagem na profundidade de preparo do solo em função da declividade do terreno. *Cerne,* 18, 607–612.

Pérez-Ruiz, M., González-De-Santos, P., Ribeiro, A., Fernández-Quintanilla, C., Peruzzi, A., Vieri, M., Tomic, S. & Agüera, J. 2015. Highlights and preliminary results for autonomous crop protection. *Computers and Electronics in Agriculture,* 110, 150–161.

Prats, S., Serpa, D., Santos, L. & Keizer, J. J. 2023. Effects of forest residue mulching on organic matter and nutrient exports after wildfire in North-Central Portugal. *Science of the Total Environment,* 885, 163825.

Programa Cooperativo Sobre Mecanização E Automação Florestal 2021. *Mecanização e Automação da Silvicultura.* Piracicaba: IPEF.

Ramantswana, M. 2023. An assessment of earth auger pitting and backpack planting productivity and quality. *In:* Report, P. C. (ed.). Ixopo: Forsilvitech, 1–30.

Ramantswana, M., Brink, M., Little, K., Spinelli, R. & chirwa, P. 2021. Perspectives and drivers of modernisation of silviculture re-establishment in South Africa. *Southern Forests: A Journal of Forest Science,* 83, 79–87.

Ramantswana, M. M., Brink, M. P., Little, K. M., Spinelli, R. & Chirwa, P. W. 2020. Current status of technology-use for plantation re-establishment in South Africa. *Southern Forests: A Journal of Forest Science,* 82, 313–323.

Reichert, J. M., Morales, C. A. S., Lima, E. M., de Bastos, F., Sampietro, J. A., de Araújo, E. F. & Srinivasan, R. 2021. Best tillage practices for early-growth of clonal eucalyptus in soils with distinct granulometry, drainage and profile depth. *Soil and Tillage Research,* 212, 105038.

Roberts, J., Little, K. & Rolando, C. 2021. Estimated herbicide use in the commercial forest sector in South Africa. *Australian Forestry,* 84, 108–121.

Roberts, J. C., Little, K. M. & Light, M. E. 2016. The use of glyphosate for the management of secondary coppice regrowth in a Eucalyptus grandis× E. urophylla coppice stand in Zululand, South Africa. *Southern Forests: A Journal of Forest Science,* 78, 217–223.

Roberts, J. C., Little, K. M. & Light, M. E. 2017. Interaction between *Eucalyptus grandis x E. urophylla* coppice reduction heights to single stems and secondary coppice regrowth in Zululand, South Africa. *ICFR Technical Note No. 03/2017.* Pietermaritzburg: Institute for Commercial Forestry Research.

Rocha, J. H. T., De Moraes Gonçalves, J. L., Gava, J. L., De Oliveira Godinho, T., Melo, E. A., Bazani, J. H., Hubner, A., Junior, J. C. A. & Wichert, M. P. 2016. Forest residue maintenance increased the wood productivity of a Eucalyptus plantation over two short rotations. *Forest Ecology and Management,* 379, 1–10.

Roland, C., Little, K. & Norris, C. 2008. *Results from a Pinus Patula Trial Investigating Combined Planting and Harvest Residue Management Options*. Pietermaritzburg: Institute for Commercial Forestry Research (ICFR).

Rolando, C. & Little, K. 2006. *The Effect of Different Methods of Harvest Residue Management on Pit Quality during Pine Re-establishment*. Pietermaritzburg: Institute for Commercial Forestry Research.

Rolando, C. A., Baillie, B. R., Thompson, D. G. & Little, K. M. 2017. The risks associated with glyphosate-based herbicide use in planted forests. *Forests,* 8, 208.

Salomão, P. E. A., Ferro, A. M. S. & Ruas, W. F. 2020. Herbicidas no Brasil: um breve revisão. *Research, Society and Development,* 9, e32921990.

São José, J. F. B. D., Vargas, L. K., Bayer, C., Lisboa, B. B. & Araújo, E. F. D. 2020. Initial growth and nutrition of eucalyptus under different management of harvest residues. *Floresta e Ambiente,* 27, e20180161.

Savill, P., Evans, J., Auclair, D. & Falck, J. 1997. Plantation Silviculture in Europe. Oxford: Oxford University Press.

Sciences, N. C. S. U. A. C. O. A. A. L. 2005. *North Carolina Agricultural Chemicals Manual*. North Carolina, North Carolina State University, Division of Continuing Education.

Sereghetti, G. C. 2016. *Qualidade do plantio manual e mecanizado para eucalipto e pinus*. Doctoral thesis, Universidade Estadual Paulista (UNESP).

Silva, J. R. V. D., Alves, P. L. D. C. A. & Toledo, R. E. B. D. 2012. Weed control strip influences the initial growth of Eucalyptus grandis. *Acta Scientiarum. Agronomy,* 34, 29–35.

Smith, C., Little, K. & Norris, C. 2001. The effect of land preparation at re-establishment on the productivity of fast growing hardwoods in South Africa. *Australian Forestry,* 64, 165–174.

Smith, C. W. 2012. Sustainability of forest plantations. *In:* Bredenkamp, B. & Upfold, S. (eds.) *South African Forestry Handboook*. 5th ed. Pretoria: South African Institute of Forestry, 279–288.

Smith, D. M., Larson, B. C., Kelty, M. J., Ashton, P. M. S. & Bowersox, T. 1997. *The Practice of Silviculture: Applied Forest Ecology*. John Wiley & Sons, Incorporated.

Smith, R. G. B., Glencross, K., Nichols, J. D., Grant, J. C. & Sethy, M. 2012. A review of site-preparation, fertilizer and weeding practices for tropical plantation species with recommendations for whitewood (Endospermum medullosum) in Vanuatu. *International Forestry Review,* 14, 433–441.

Soler, R. R., Guerra, S. P. S., Esperancini, M. S. T., Sereghetti, G. C. & Oguri, G. 2019. Avaliação econômica da operação de plantio mecanizado de eucalipto em dois diferentes espaçamentos. *Energia na Agricultura,* 34, 462–470.

Soler, R. R., Guerra, S. P. S., Oguri, G. & Rodrigues, S. A. 2024. Economic evaluation and performance of a tree planting machine performing in two different slope classes and conditions of harvesting slash. *Croatian Journal of Forest Engineering: Journal for Theory and Application of Forestry Engineering,* 45, 15–24.

Solgi, A., Naghdi, R., Zenner, E. K., Hemmati, V., Behjou, F. K. & Masumian, A. 2021. Evaluating the effectiveness of mulching for reducing soil erosion in cut slope and fill slope of forest roads in Hyrcanian Forests. *Croatian Journal of Forest Engineering: Journal for Theory and Application of Forestry Engineering,* 42, 259–268.

Souza, A. 2002. Preparo de solos coesos para cultura do eucalipto no extremo sul da Bahia. *In:* Gonçalves, J. L. M. & Stape, J. L. (eds.) *Conservação e cultivo de solos para plantações florestais*. Piracicaba: IPEF, 297–311.

Souza AJ, A. E., Silva AP, 2001. Preparo de um solo coeso para cultura do eucalipto no extremo sul da Bahia. *Anais Congresso Brasileiro de Ciência do Solo*. Bahia.

Sullivan, T. P., Sullivan, D. S. & Klenner, W. 2021. Fate of postharvest woody debris, mammal habitat, and alternative management of forest residues on clearcuts: a synthesis. *Forests,* 12, 551.

Swift, R. 2020. *Cutting Down and Mulching Trees: Disc Mulcher vs. Drum Mulcher* [Online]. Online: Diamond mowers. Available: https://blog.diamondmowers.com/disc-mulcher-vs-drum-mulcher [Accessed 14 March 2024 2024].

Timber Pesticide Working Group. 2023. *Standard Operating Procedure Aerial Spraying – SANS 10118*. Online: TIPWG.

Titshall, L., Dovey, S. & Rietz, D. 2013. A review of management impacts on the soil productivity of South African commercial forestry plantations and the implications for multiple-rotation productivity. *Southern Forests: A Journal of Forest Science,* 75, 169–183.

Torresan, C., Berton, A., Carotenuto, F., DI Gennaro, S. F., Gioli, B., Matese, A., Miglietta, F., Vagnoli, C., Zaldei, A. & Wallace, L. 2017. Forestry applications of UAVs in Europe: A review. *International Journal of Remote Sensing,* 38, 2427–2447.

USDA Forest Service, R. M. R. S. 2004. *Fuels Planning: Science Synthesis and Integration; Economic Uses Fact Sheet 01: Mastication Treatments and Costs*. Fort Collins, CO: Rocky Mountain Research Station.

Velusamy, P., Rajendran, S., Mahendran, R. K., Naseer, S., Shafiq, M. & Choi, J.-G. 2021. Unmanned Aerial Vehicles (UAV) in precision agriculture: Applications and challenges. *Energies,* 15, 217.

Viero, P. 2004. *Eucalypt Establishment and Re-establishment*. Pietermaritzburg: Institute for commercial Forestry research.

Viero, P. 2007. *Planting Eucalypts using Hydrogels*. Pietermaritzburg: Institue for Commercial Forestry Research.
Viero, P. & Du Toit, B. 2012. Establishment and regeneration of eucalypt, pine and wattle stands. In*:* Bredenkamp, B. & Upfold, S. (eds.) *South African Forestry Handbook*. 5th ed. Pretoria: Southern African Institute of Forestry.
Viero, P. W. & Little, K. M. 2006. A comparison of different planting methods, including hydrogels, and their effect on eucalypt survival and initial growth in South Africa. *Southern African Forestry Journal,* 2006, 5–13.
Wang, L., Huang, X., Li, W., Yan, K., Han, Y., Zhang, Y., Pawlowski, L. & Lan, Y. 2022. Progress in agricultural unmanned aerial vehicles (UAVs) applied in China and prospects for Poland. *Agriculture,* 12, 397.
Wilcken, C., Lima, A., Dias, T., Mateus, G. & Guerra, S. 2011. *Guia prático de manejo de plantações de eucalipto*. Botucatu: FEPAF.
Zitha, H. M., Ramantswana, M. M. & Spinelli, R. 2023. Performance of harvesting residue treatment methods in South African pine plantations. *Forests,* 14, 1661.
Zulu, T. 2023. *Comparing the productivity of pitting machine operators on different sites 1 conditions*. Honors Natural Resource Management Research Project, Nelson Mandela University.
Zwane, B. 2023. Early lessons on mechanized planting in Eswatini. *In:* Ramantswana, M. (ed.) *Modern Silviculture Webinar – Optimising Modern Silviculture* [Online]. George: Nelson Mandela University. Available: https://www.youtube.com/watch?v=y_-6znzOU34.
Zwolinski, J. & Groenewald, W. 2004. Natural regeneration of pine plantations in South Africa as a cost-effective way of stand re-establishment. *Forestry,* 77, 483–493.
Zwolinski, J., South, D., Cunningham, L. & Christie, S. 1996. Weed control and large bare-root stock improve early growth of Pinus radiata in South Africa. *New Zealand Journal of Forestry Science,* 26, 163–172.

4 Silviculture technologies in stand tending

Muedanyi M. Ramantswana and Saulo P. S. Guerra

4.1 Introduction

This chapter will describe the application of silviculture technologies in stand tending, the management phase crucial for facilitating optimal tree growth, health, and timber value. This chapter will explore various activities, including weeding after planting, forest protection, pruning, and thinning, while emphasising the importance of timing to maximise resource investment and prepare stands to yield optimal volume at harvesting. The reader will learn about various stand-tending technologies, from basic manual techniques to mechanised and advanced methods. A comprehensive understanding of stand tending will facilitate optimal returns and reduce risks to volume losses in the short and long term.

4.2 Weeding after planting

Weeding refers to removing unwanted vegetation (e.g. grass, herbs, shrubs and coppice), which can suppress and affect the trees in the early growth stage (Park et al., 2020). Weeding is an essential activity after the establishment phase and must be conducted promptly until the trees reach canopy closure (Evans and Turnbull, 2004, Pereira et al., 2012). Weeds growing amongst newly established plants compete for resources (light, nutrients, water), thereby restricting the plant growth (Vasic et al., 2012, Silva et al., 2012). The composition of weeds in a specific stand depends on climatic conditions (temperature and precipitation) and the infestation through natural migration or the introduction of artificial species.

Various methods, from manual, motor manual, semi-mechanised, mechanised, and aerial technologies, can remove weeds within a growing stand (Byrnes et al., 1973). In South Africa, weeding after planting is considered the least technologically advanced or modernised activity compared to other re-establishment activities (Ramantswana et al., 2021). In Brazil, the implements used for applying herbicides between the planting rows have evolved to increase efficacy and efficiency. Out of the total weeded area of 127,674 hectares weeded, most of the weeding was conducted using chemical spraying

DOI: 10.4324/9781003391227-4

with a backpack (43.8%) followed by manual slashing (22%), manual ring weeding (17.2%), tractor boom sprayer (7.3%), wind boxes (2.7%) and other methods (7.1%) (Ramantswana et al., 2020). In Brazil, this type of survey was not carried out. This book does not focus on the use of covering plants, grazing, and biological control measures to limit the growth of competitive weeds.

Burnt sites tend to produce more vigorous weeds than sites with residues (post-harvest or mulched) covering the surface (Lantagne and Burger, 1987); therefore, more regular assessments and weeding are required. This is important within the South African context, where most harvest residues are burnt after planting (Ramantswana et al., 2020). Furthermore, it is important to complement any fertilisation programmes with good weed control to facilitate optimal tree growth and derive the required nutritional benefits to the plant (Botha et al., 2012). The way weed control is conducted is directly influenced by the species, site, and climate conditions prevalent in the area. The extent to which weeds are removed depends on the weeding pattern followed (Figure 4.1). For optimal growth, the complete weeding pattern is preferred; however, the cost of resources used and potential environmental impacts (e.g. erosion) need to be considered. Sometimes, a hybridised pattern may be more feasible.

4.2.1 Manual method

Manual weeding involves using infield labour with tools (e.g. hoe and slasher) to remove and clear weeds in a 50–75-cm weed control strip for reforestation

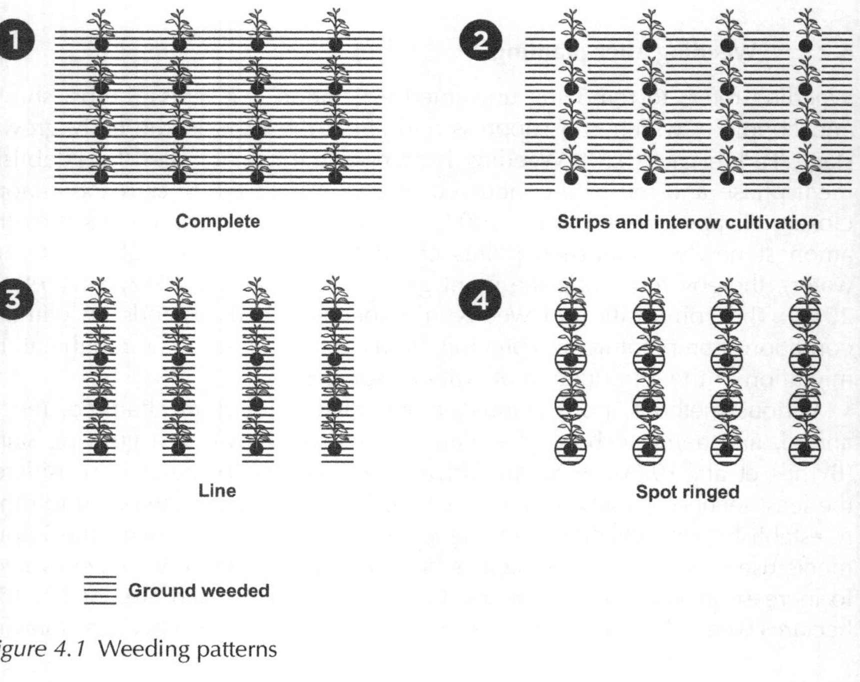

Figure 4.1 Weeding patterns

(Silva et al., 2012). Manual weeding can sometimes be conducted without tools, such as by hand pulling weeds. Manual weeding is labour-intensive, has poor ergonomics, is slow, and is a costly activity with limited effectiveness (temporary cutback). Manual control of weeds can be economically challenging or unfeasible because the weed seed bank remains on the topsoil (Gonçalves et al., 2008) and may not effectively control invasive plants that can regenerate when cut (Salomão et al., 2020). However, it requires very little capital investment, is flexible (e.g. it can work in different terrain conditions), and can be adapted to various operational conditions (Lopes et al., 2011), such as steep terrains or having many stumps or termite mounds.

4.2.1.1 Ring weed

Ring weeding uses a hoe to clear most weeds around the tree. The weeds closest (20 cm) to the planted tree are removed by hand pulling, and the remaining ones are cleared using a hoe. The cleared perimeter of the ring weed must be 1 m around each planted tree. When a hoe is used for removing weeds, the soil may be displaced from around the trees, and this soil needs to be carefully redistributed before proceeding to the next plant.

4.2.1.2 Slashing

Slashing is carried out using a slasher tool. A slasher is an implement with a long sharp blade used to cut weeds. The weeds growing in the line can be cleared. The infield labourers must keep a safe working distance from each other to prevent any injuries. A slashing operation is effective in clear stands overgrown with grass by lowering the vegetation to a level that is more accessible for additional treatment methods. Any slashing operation must be followed up with a chemical application.

4.2.2 Motor manual method

Motor manual weeding involves the use of a motorised tool like a brush cutter or a clearing saw to clear any weeds within the stand. A brush cutter is an outdoor tool for cutting grass and herbaceous weeds (Ackerman et al., 2018). A brush cutter can be used with cutting attachments like a wirehead, a 2-tooth blade, a 3-tooth blade, and a 24-tooth blade (Naskrent et al., 2020). This method requires an investment in tools, personal protective equipment, and training.

Brush cutters are designed to cut light herbaceous weeds, woody herbaceous weeds, reeds and woody bushes, heavy bushes, and thin trunks; however, the effects are not prolonged as regrowth can reestablish quickly if not treated with herbicide (Mallik et al., 2002). For maintenance operations in plantations, the strip cutting and release cutting methods can be applied. Strip cutting refers to a working technique where the operator moves between the

Figure 4.2 Motor manual cutting around and along the rows

trees in up and down paths, cutting them alternately. Release cutting is a technique where the operator only clears vegetation around the plant (Figure 4.2) (Ackerman et al., 2018).

4.2.3 Mechanical method

Mechanical weeding methods involve the use of equipment such as a tractor with an implement to cut or dig up and mix the weeds into the soil. Implements that can slash/mow, disc, flail, and mulch can be used to clear the unwanted vegetation within the planted stand. The implements can be attached to the drawing vehicle, like a tractor, using the three-point linkage and driven by the power take-off (PTO) (Figure 4.3). When applying mechanical weeding methods, care must be taken to avoid causing physical damage to the trees. Mechanical methods are capital-intensive (compared to manual methods) and limited to working in the line or interrow where the carrier vehicle can access. Furthermore, mechanical equipment may be restricted by terrain conditions (ground conditions, ground roughness, slope) (Ehrentraut and Branter, 1990, Rajendran et al., 2014), high embankments, residues in the stand (stumps, branches, logs) and tree spacing. Mechanical operations such as slashing, mowing, and flailing are more productive compared to manual methods.

4.2.3.1 Mechanical slasher/mower/flail

Mechanical weeding can be performed using a rotary mower or a flail mower. Both attachments can be used to clear and cut grass, brush, and dense vegetation within the stand. A rotary mower uses an implement with blades that

Figure 4.3 Discing between trees with a tractor pulling a plough

rotate, thereby cutting the unwanted vegetation within the interrow (Swift, 2020). The flail mower comprises Y or T-shaped knives that rotate on the axle, clearing any vegetation (e.g. grass, plants and brush) as the flail moves parallel to the ground (Swift, 2020). Flail mowers can be PTO or hydraulically driven, while rotary mowers are usually hydraulically powered. As the mover moves, the cut vegetation is broadcast behind the implement. The process of mowing removes weeds that take up space and light resources, but this type of weeding requires repetition, especially during the growing season (Vasic et al., 2012). It is recommended that any slashing operation be followed by a chemical application to kill the regrowth effectively.

4.2.3.2 Mechanical disc harrowing

Disc harrowing is the process where a disc plough is used to uproot, cover and damage weeds (Peruzzi et al., 2017) in between the interrows for vegetation management and fuel load reduction. Studies have shown that disc harrowing activity has no significant impact on below-ground tree growth (Madeira et al., 2012, Goldsmith, 2019). The disturbance caused by the harrow tines can be adjusted by altering the angle of the tines with reference to the ground conditions.

4.2.3.3 Mechanical mulching

Mulching is a method that can be applied to cut, chop, and grind weeds growing in the interrow. Mulching as a weed control measure is usually carried out in conjunction with residue or fuel load reduction. For details on mulcher configurations and their application in residue management, refer to Chapter 3, Section 3.2.2. Mulching as a weeding after planting is typically conducted using a small tractor-drawn mulching implement. Smaller equipment can fit into the narrow interrow without damaging the young stand. Mulching operations must be followed by chemical application when the weeds regrow through the mulch. The mulched material can affect the soil temperature and enhance physical, chemical and biological properties of the soil (Monteiro and Santos, 2022). Interestingly, Deng et al. (2020) compared mechanical mulching with chemical weed control in eucalyptus stands and found that both methods effectively removed weeds; moreover, after one month, there were no significant differences in tree growth between the two methods.

4.2.4 Chemical method

Chemical weed control involves the application of herbicides on weeds or pesticides on trees. Herbicides can be selective or non-selective. Selective herbicides are applied to kill specific weeds and not kill a particular plant deemed valuable, and non-selective (broad-spectrum) herbicides kill all vegetation within a specific target area (Oregon State University, 2024). The application rate and the application method influence the application of herbicides. According to Miller and Mitchell (1990), herbicides can be classified as follows:

- Foliar active – applied as a spray or mist to leaves and twigs where infiltration occurs
- Soil active – applied as a liquid or a granule and absorbed through the root system. Requires precipitation for it to dissolve and then be taken up by the roots
- Systematic – the process of how the tree is affected by the chemical after uptake through the leaves or roots. Most forest herbicides are systematically disrupting the physiological processes in the weeds
- Contact – only affects the plant surfaces that come in direct contact with the herbicide. It may not kill the whole plant
- Woody control – only effective in controlling woody plants
- Herbaceous control – only effective in controlling broadleaf plants, e.g. grasses

Chemical control is the predominant method and standard for weed control in South Africa and Brazil (Donald, 1986). Most chemical applications (glyphosate-based and metazachlor products) are applied during the re-establishment phase (before canopy), as this is the period when the stand is most exposed to competition and young plants are most vulnerable (Ndlovu et al., 2022, Roberts et al., 2021). Chemical application is known to be better than manual and mechanical methods because it provides prolonged weed suppression, as the weeds are killed, unlike other methods, which only cut back the weeds.

Certification bodies require forestry organisations to use integrated pest management to reduce and eliminate the use of chemical pesticides (Forest Stewardship Council, 2019). The use of glyphosate when using knapsacks and tractor-powered systems is considered relatively safe, with the use of personal protective equipment (Machado-Neto et al., 2000).

The extent of chemical weeding after planting can differ for softwood and hardwood species; however, some of the common spray operations include: broadcast spray, row spray, spot spray, secondary coppice regrowth spray, pre-pruning spray, pre-thinning spray, post-thinning spray, and pre-harvest spray. After canopy closure, most chemical applications are target-specific (Roberts et al., 2021). Refer to Annexure 1 for approved chemicals used in the South African forestry industry (TIPWG, 2024).

4.2.4.1 Knapsack/backpack sprayer

A manual sprayer has dimensions of 51 × 35 × 19 cm, weighs 5–7 kg when empty, and holds 12–25 litres of product and water, resulting in a total weight of over 30 kg. Knapsack or backpack sprayers are the most common tools used for chemical application in forests because they are inexpensive and adaptable to different operating conditions (Lopes et al., 2011, Roberts et al., 2021). However, when trees are young, there is a risk of spraying or scorching them during the spraying operation. Therefore, when chemical spraying is executed amongst young trees, they are covered with cones to prevent any chemical drifting and killing them. Overall, chemicals must be applied consistently on the site to avoid under- or over-application, which may result in chemical waste or poor results. When applying chemicals, the spray droplet composition and area coverage are essential to ensure the required weed-killing effect is achieved when using knapsack sprayers (Teka et al., 2012).

a Manual knapsack

Manual knapsack spraying after planting can be executed as a full-cover spray or as a spot spray. A full-cover spray is commonly used when weeds cover the entire surface of the intra- and interrow, or after a mechanical weeding operation (Figure 4.4). The manual knapsack is dependent on the manual pumping of the chemical mix by the operator. When applying chemicals using a knapsack, it is essential to ensure that it is calibrated to the chemical application needed to kill the weeds effectively. The work cycle involves several phases: product application, empty travel, refilling, mixing, full travel, maintenance, cleaning, and breaks.

b Motorised knapsack

The motorised knapsack sprayer runs on a two-stroke engine. If features similar parts to the manual knapsack but it has improved ergonomics and power, which make it an efficient tool to work with. It can be used for pre-plant, full cover and selective herbicide spraying work in forestry. It delivers a consistent spray pressure compared to a manual knapsack, where manual pumping reduces efficacy and productivity. Droplet size can be adjusted by varying the pressure of the sprayer: a higher pressure reduces droplet size, while a lower pressure increases it.

Figure 4.4 Knapsack spraying after planting

4.2.4.2 Ground-based vehicle sprayers

Chemical application after planting can be performed with a tractor, dozer, skidder, or a suitable ground-based machine with a spraying rig. Selective chemicals are used after planting to avoid damaging the valuable crops. Ground-based vehicle sprayers can apply herbicides that are both foliar and soil active. The tractor system needs to access the stand interrow to spray effectively.

Ground-based primary moving vehicles like tractors can be configured with a tractor-mounted boom sprayer, tractor-towed boom sprayer, tractor with lances, or tractor with wind box. When using a tractor-type system for spraying, the operator and the infield labour must be trained and skilled in using herbicides (e.g. calibration, application rates, and toxicity). Employee training should include techniques for applying chemicals, increasing operational efficiency, and ensuring even and effective coverage while minimising costs, waste and environmental impact (Vuaden Junior, 2019). The modern spraying system typically comprises safety features (power take-off guard, belt and pulley guards), a tank, PTO, electric or hydraulic pump, filters, regulator, strainer, regulator, gauge, spray boom, and spray nozzles (Miller and Mitchell, 1990, Dornelles et al., 2011, Cornago Junior, 2020, Barbaresco et al., 2021).

A shielded spray boom in Brazil is a mechanised spraying technology for drift reduction and protecting eucalypt seedlings. Shielded spray booms potentially mitigate this by providing a physical barrier that reduces the dispersal of droplets, keeping the herbicide application limited to the equipment's width (Carvalho et al., 2020). This spraying technique allows for more precise application in sensitive areas or between rows, almost eliminating the risk of off-target damage and the environmental impact of chemical weeding (Santos and Reichert, 2022).

4.2.4.3 Remotely piloted aircraft system

Since 2010, the use of remotely piloted aircraft (RPA) systems in forestry applications has increased globally. For details about unmanned aerial vehicles (UAVs), refer to Section 3.4.3. UAVs like drones were initially radio-controlled devices operated by a pilot remotely. However, drones are shifting to more fully automatic devices using GPS-based autopilot systems (Hafeez et al., 2023). The applications have mainly been limited to inventory and forest parameter estimation, conservation of natural ecosystems, pest and disease mapping, fire monitoring, and post-harvesting assessments (Eugenio et al., 2020).

In Brazil, RPAs were regulated by Ordinance No. 298, dated September 22, 2021, where the Ministry of Agriculture, Livestock and Supply – MAPA (Brasil 2021) established the rules for the operation of these aircraft for the application of pesticides and related products, adjuvants, fertilisers, inoculants, correctives, and seeds. These rules define that operators need to be registered with MAPA, have certification of completion of the Remote Aerial Application Course (CAAR), and ensure that the RPAs are in regular status with the National Civil Aviation Agency (ANAC). Additionally, for operational safety purposes, aerial applications with RPA are restricted to the target area of intervention, and it is of utmost importance to record data related to each application of pesticides and related products, adjuvants, fertilisers, inoculants, correctives, and seeds (Brasil, 2021).

In South Africa, the use of UAVs for chemical applications is still limited due to the lack of selective chemicals approved for aerial spraying. The prospects of this technology being adopted are good because:

- Good access to difficult-to-access areas
- Cost-effectiveness of the technology compared to conventional aircraft
- It could be complementary to traditional methods – aerial and terrestrial sprayers
- Improved spraying productivity
- More effective control in combating pests and diseases
- Decrease spray losses to the soil
- Elimination of soil compaction
- Improve health and safety compared to the direct in-person chemical application, e.g. backpack sprayers

Chemical spraying drones are designed to carry tanks containing chemicals and nozzles controlled by independent solenoid valves that apply chemicals when needed in conjunction with a pump. The tanks' capacities can vary from 250 ml to over 40 litres and can be coupled to flat fans, micro air, or universal nozzles (Carvalho et al., 2020) (Programa Cooperativo sobre Mecanização e Automação Florestal, 2022)

4.2.4.4 Helicopter and fixed wing

When applying pesticides to eradicate pests in forestry, aerial application of solid or liquid material is more effective compared to ground-based application methods. Aerial application can cover large areas cost-effectively and apply pesticides more efficiently (Brockerhoff et al., 2010). Fixed-wing aerial applications can cover large areas cost-effectively and efficiently due to their higher flight speeds and altitudes. Meanwhile, helicopters are more versatile and suitable when there is no runway for take-off and landing, and lower speeds and crop proximity are required (Brockerhoff et al., 2010, Jones, 2019, da Silva Cavalcante et al., 2022).

4.3 Pruning

Pruning is a wood quality improvement operation wherein branches from a tree stem are cut off to produce knot-free timber, also known as clearwood. It is necessary when the objective is a final product for sawnwood or veneer production (Scolforo et al., 2001, Hoppe and Freddo, 2003, Evans and Turnbull, 2004, Fritscher, 2022). Furthermore, pruning can be used for fire protection (access and fuel load management) (Brink, 2012), and providing better access within the forest stands for other silvicultural operations (Lopes et al., 2013). Pruning is necessary for both the access of machines and equipment in operational areas and access to firebreaks. This essential activity is still hazardous, mostly done manually, exposing workers to a high risk of accidents (Programa Cooperativo sobre Mecanização e Automação Florestal, 2022). For this purpose, mechanised circular blades, mounted on a 4x2 FWD tractor, could cut brushes and branches from 5 to 13 cm in diameter (Figure 4.5).

Figure 4.5 Mechanised pruner

Even though some eucalypts are self-pruning (dead branches shed off and occlusion of wounds and stubs as the trees grow), timely pruning can help prevent loose and poor-quality timber due to knots. Eucalypt pruning is not a silvicultural operation in Brazilian plantations, but only in pine plantations.

Live and dead branches can be pruned from a tree; however, pruning live branches is beneficial in managing the knotty core (Montagu et al., 2003). Pruning has an impact on physiological characteristics (photosynthesis and leaf area index) and responses (Pinkard and Beadle, 2000), being economically viable due to improved wood quality, and potentially higher prices (de Oliveira et al., 2012). In South Africa and Brazil, there are two methods of pruning: manual pruning with a pole saw and motor manual pruning with a pruning saw (Sibiya et al., 2021, Lopes et al., 2013). Other options, such as electric shears, have been tested, but the technology has not been adopted despite the positive results in South Africa (Shekwa et al., 2017). In Brazil, electric shears were evaluated in 21-month-old eucalypt plantations, and they could be up to 62% more productive (132 trees pruned per hour) than manual pruning (Programa Cooperativo sobre Mecanização e Automação Florestal, 2020). The average heartbeats per minute (bpm) were 11% higher during manual pruning with shears than when using electric pruning.

No fully mechanised pruning technology is available in either country. The most appropriate pruning method usually depends on managing the cost and benefits of clear wood and the cost to attain it (Montagu et al., 2003). Standard pruning heights vary from 2 to 7 m (Warkotsch, 1988), but some may reach up to 10 m (Table 4.1). This operation cannot exceed 50% of the tree's height or 1/3 of its live crown (Scolforo et al., 2001). Higher pruning is more expensive, less productive, and riskier, as the pruner may require a ladder to reach the required pruning height.

4.3.1 Manual pruning

Manual pruning is the predominant method used for pruning pine and eucalyptus stands in South Africa and Brazil (Figure 4.6). The branch is cut flush to the stem using a pruning curved blade saw (Lopes et al., 2013). The pruning

Table 4.1 Typical pruning schedule (Evans and Turnbull, 2004)

Pruning operation	*Pruned height (m)*	*Approximate timing*	
		Stand height	*Stand condition*
First (low pruning)	2.5	6.0	Just after canopy closure
Second (high pruning)	5.0	9.0	Before the first thinning
Third (high pruning)	7.5	12.0	At the time of the first thinning
Fourth (high pruning)	10.0	15.0	Before the second thinning

Figure 4.6 Manual pruning

height can be adjusted by replacing the pole attached to the pruning saw, removing branches with diameters ranging from 2 to 15 cm. Keeping the pruning saw blade sharp is essential for efficient cutting and to avoid tree damage. When performing manual pruning at 2 m and 3.5 m, a person can use a billhook to prune an average of 34–66 trees per scheduled machine hour (Sibiya et al., 2021, Ferreira et al., 2019). Manual pruning is costly, demands more significant physical effort, and the time to prune each tree is directly influenced by pruning height, number of branches per tree, maximum branch diameter, and number of trees to be pruned per hectare (Evans and Turnbull, 2004, Shekwa et al., 2017). However, manual pruning can offer better quality in avoiding injuries to the tree´s wood compared to semi-mechanised methods (Ferreira et al., 2019).

4.3.2 Motor manual pruning

A pole pruner is a motorised tool designed to cut tree branches from the tree stem (Figure 4.7). The operator uses a tool supported by a backpack harness,

Figure 4.7 Motor manual pruning

and the actual machine comprises components such as a chainsaw linked to a drive shaft and powered by a motor (Ackerman et al., 2018). Different models with varying lengths and weights are used, depending on the pruning height (de Oliveira et al., 2012). When performing motor manual pruning at 2 m and 3.5 m, an operator can prune an average of 71–136 trees per scheduled machine hour (Sibiya et al., 2021). Furthermore, when motor manual pruning is compared to manual pruning in pine pruning operations, it is more ergonomically (specifically, discomfort) friendly, even though the machine to be carried could weigh up to 9.0 kg (de Oliveira et al., 2012, Lopes et al., 2013).

4.4 Forest protection

The scope of this section focuses on the technologies available to protect forests against biotic and abiotic agents, which can have a devastating impact on plantation forests. Plantation forest trees are vulnerable to stress and damage by biotic factors such as pests and diseases, as well as abiotic factors (e.g. fire, wind, frost, hail, snow, drought and other natural phenomena) (Gadgil and Bain, 1999, Robinne, 2021). Biotic and abiotic factors can be identified through their characteristics. Biotic factors are living (e.g. fungi, bacteria, insects), are host-specific, can be infectious, and can often impact the forest stand randomly. Abiotic factors (wildfires, droughts) are not specific to a host and are non-infectious, and damage is expressed similarly across the affected stand (Roux et al., 2012).

Although inferences are made regarding prominent biotic (diseases and pathogens) and abiotic factors, which pose risks to the forestry industries in Brazil and South Africa, extensive details about these threats are beyond the scope of this book, and other comprehensive relevant literature can be consulted. This chapter will focus on technologies that are operationally available to monitor and protect plantation forests from pests and diseases, as well as wildfires, as these are becoming increasingly important (Payn et al., 2015). Operational tools and technologies for applying pesticides to treat pests and diseases have been described under Section 4.1.4.

4.4.1 Pests and disease protection

The forestry industry recognises the risks and likelihood of emerging pests and diseases. Pests and diseases that affect plantation forestry species can originate from their native regions or from local species where the exotic trees have been introduced (host shift) (Wingfield et al., 2008). Protecting plantation forest trees can be achieved through three main control strategies: prevention (quarantine), eradication (early detection and implementation of effective control measures), and management (including chemical and biological control, as well as breeding and selection) (Chungu et al., 2010). According to Wingfield et al. (2001), strategic measures to reduce plantation risks from pests and diseases include diversifying planting stock and increasing knowledge of pests and pathogens. Furthermore, South African forestry growers have diversified their species, with a notable rise in pine and eucalypt hybrids (Morris, 2022), as is the case in Brazil, according to (Programa Cooperativo sobre Mecanização e Automação Florestal, 2025), including a cooperative breeding programme for the Corymbia genus.

In South Africa, pesticides and their quantities are strictly regulated according to legislation. Furthermore, contractors who handle pesticides and offer services involving the use of agricultural remedies need to be registered as a pest control operator (PCO). The use and application of pesticides need to be considered in conjunction with the holistic integrated pest management framework. Forest protection technologies used for eradication using pesticides are similar to technologies used for chemical stand tending after planting. The composition and hazardous nature of pesticides differ, but the application technologies and tools are relatively the same. Pesticides can be applied manually with a backpack sprayer, mist blower, UAVs, and aircraft.

4.4.1.1 Diseases

Plantation species are susceptible to attacks by diseases, and once infected, severe losses can be experienced (Wingfield, 1990). Diseases can affect the roots, stems, branches, or leaves of plantation forestry species. Some of the prominent diseases that affect plantation forestry species in Brazil and South Africa are listed in Table 4.2.

Table 4.2 Plantation forestry diseases (Roux et al., 2012)

Disease name	*Genus affected*	*Tree parts affected*	*Brazil*	*South Africa*
Armillaria root rot	E, P, A	Roots and root collar	√	√
Aulographina leaf spot	E	Leaves and petioles	√	√
Bacterial blight	E	Leaves		√
Bacterial wilt	E	Roots		√
Botryosphaeria canker	E, A	Stems, branches	√	√
Calonectria leaf blight	E	Leaves	√	√
Camptomerris leaf spot	A	Leaves		√
Ceratocystis/wattle wilt	A, E	Stems/vascular system	√	√
Cercoseptoria needle disease	P	Needles	√	√
Chrysoporthe canker	E	Root/collar, stems		√
Coniothyrium/measles canker	E	Stems, branches	√	√
Cyclaneusma needle spot	P	Needles		√
Cylindrocladium pteridis	P	Needles	√	
Diplodia diseases	P, E	Stems, branches, roots, sawtimber	√	√
Dothistroma needle blight	P	Needles		√
Elsinoe masingae	E	Leaves		√
Fusarium spp.	P	Roots	√	
Kirramyces leaf spot	E	Leaves		√
Lophodermium needle cast	P	Needles		√
Mildew	A,E	Leaves		√
Mycosphaerella leaf blotch diseases	E, P	Leaves	√	√
Myrtle rust – *Austropuccinia psidii*	E	Leaves, shoots, stem	√	
Phytophthora root rot	A, E, P	Roots, root collar	√	√
Pink disease	A, E	Stems and branches	√	√
Pitch canker	P, E	Stems, branches, cones	√	√
Pseudophaeolus root rot	R	Roots		√
Quambalaria leaf and shoot blight	E	Leaves, shoots, stems, branches	√	√
Rhizina root disease	P	Roots	√	√
Sooty mould	P	Branches and needles	√	
Teratosphaeria leaf blight	E	Leaves	√	
Tip blight	P	Branches and needles	√	
Valsa (Cytospora) canker	E, P	Stems, branches	√	√
Wattle leaf rust	A	Leaves		√

4.4.1.2 Insect pests

Some of the prominent diseases that affect plantation forestry species in Brazil and South Africa are provided in Table 4.3.

4.4.1.3 Pest and disease technologies

When working with pests and diseases, it is essential to use an integrated pest management process. The success of pests and diseases depends on intricate

Table 4.3 Insect pests in plantation forestry trees

Disease name	*Genus affected*	*Feeding*	*Brazil*	*South Africa*
Ambrosia beetles	P, E	Wood borer	√	√
Black pine aphid	P	Sap-sucker		√
Bronze bug	E	Sap-sucker defoliator	√	√
Brown lappet moth	A, E, P	Defoliator	√	√
Catamonus weevil	A	Defoliator		√
Cossid moth	E nitens	Wood borer		√
Cutworms	A, E, P	Establishment pest		√
Deodar weevil	P	Bark borer		√
Eucalyptus tortoise beetle	E	Defoliator	√	√
Eucalyptus gall wasp	E	Gall former	√	√
Eucalyptus longhorn beetle	E	Wood borer	√	√
Eucalyptus weevil	E	Defoliator	√	√
Eucalyptus snout beetle	E	Defoliator	√	√
Grasshoppers and crickets	A, E, P	Establishment pest	√	√
Green bronze beetle	E, in gP	Establishment pest Defoliator		√
Grey weevil	E	Establishment pest Defoliator		√
Leaf-cutting ants (Atta/ *Acromyrmex* spp.)	E, P	Defoliator	√	
Lesser wattle chafer	A	Defoliator		√
Mediterranean pine engraver beetle	P	Bark borer	√	√
Pine bark beetle	P	Establishment pest Bark borer	√	√
Pine brown tail moth	P	Defoliator		√
Pine emperor moth	A, P	Defoliator		√
Pine looper	E, P	Defoliator	√	√
Pine woolly aphid	P	Sap-sucker	√	√
Poplar emperor moth	A, E, P	Defoliator		√
Red-haired pine bark beetle	P	Bark borer		√
Sirex woodwasp	P, E	Wood borer	√	√
Termites	A, P, E	Establishment pest	√	√
Wattle bagworm	A	Defoliator		√
Wattle chafer	A, P	Defoliator		√
Wattle emperor moth	A	Defoliator		√
Wattle mired	A	Sap-sucker		√
Wattle semi-looper	A	Defoliator		√
Whitegrubs	A, E, P	Establishment pest		√

Roux et al. (2012), Jordan et al. (2021).

links between factors such as tree nutrition, production activities, other pest and disease outbreaks, and sustainability concerns (Herdiyeni et al., 2017). Various technologies, such as surveillance, sensors, and information technologies, are required to manage pests and diseases in plantation forests effectively. Early detection of forest health problems can help minimise stand productivity losses (Stone and Mohammed, 2017). Although new technologies significantly improve the monitoring and detection of pests and diseases, traditional manual methods of field assessments and pest detection remain quite accurate. However, manual processes are slow, costly, and cannot be applied at a large scale. This section will cover the primary pest and disease monitoring and detection technologies.

(i) Surveillance technologies

Surveillance technology can also be referred to as monitoring tools. In forest health assessments, assessing damage location, extent, and impact is important in guiding effective control measures (Zhang et al., 2019). However, any forest health programme must be accompanied by expertise for correct diagnosis and on-ground identification of the damaging agents (Stone and Mohammed, 2017). Assessing pests and diseases in plantation forests usually depends on aerial detection surveys of the tree crowns (foliage discolouration and defoliation) (Rullan-Silva et al., 2013). Changes in the characteristics of the tree crown (reflected radiation between damaged and healthy trees) can be a good indicator of tree health based on vegetation indices.

Remote sensing technology is an effective tool for pest and disease monitoring. This technology is used to identify and collect digital imagery for pest and disease identification. According to Zhang et al. (2019), the primary remote sensing technologies include (i) Visible and near-infrared spectral systems (VIS-SWIR) – damage assessed by reflectance in the VIS-SWIR region, (ii) Fluorescence and thermal systems – capture pre-symptom physiological changes, (iii) Synthetic aperture radar and Lidar – capture structural or morphological changes caused by prevalent diseases and pests. Remote sensing can be categorised based on the characteristics of the data collected, such as spectral (forest imagery based on tree canopy information), spatial (high spatial resolution satellite and airborne imagery for extraction of image objects) and temporal (use of data archives for forest health monitoring over time) (Chen and Meentemeyer, 2016).

4.5 Wildfire protection

In South Africa, fire poses a significant risk that can cause substantial damage over a short duration. From 1980 to 2018, the South African forest industry experienced fire damage on 682,339 hectares of land, an estimated 22% of this damage occurring between 2009 and 2018 (Godsmark and Oberholzer, 2019). In Brazil, no official and reliable records are available for fires in plantation forests. However, between 2020 and 2023, an estimate suggests that between 30,000 and 70,000 hectares per year of eucalyptus and pine were consumed by fire. Fires can be caused by natural causes, such as lightning, manmade ignition, burning firebreaks, fuel load reduction operations, and arson. For a fire to occur, the core components comprising oxygen, fuel, and ignition temperature must be present (Robinne, 2021, Shikwambana et al., 2019). When fires occur, they can potentially cause damage to trees, the environment, and infrastructure, leading to health issues and loss of lives (Strydom and Savage, 2016). The extent of damage or losses can be influenced by extreme weather conditions, drought, lack of fuel load management, and the prevalence of highly combustible material (Kraaij et al., 2018). Of these factors, the only factor the forester can influence is the fuel. Different fire types, such as ground, surface, and crown fires, can occur. This section focuses on the varying levels of technology available for monitoring, preventing, and suppressing fire in plantation forests. Furthermore, the application of manual, semi-mechanised, mechanised, and aerial technologies will be described.

4.5.1 Forest fire detection technologies

Early fire detection and communication are essential for effectively monitoring and controlling fire events. The extent to which a fire burns an area is influenced by how early it is detected and the availability of resources required to suppress the fire (Den Breejen et al., 1998). Fire detection systems are responsible for identifying smoke, determining the location of the fire and communicating details related to the fire to the control room/ forester or fire authority designated for the area. According to Alkhatib (2014), detection and monitoring systems can be conducted through operational detection systems such as fire towers (manual lookouts), aerial patrols, electronic lightning detectors, and automatic detection systems (Figure 4.8). This section will focus on the use of advanced fire detection systems.

i Automated fire detection systems

This system comprises individual or ground sensors (fixed, PTZ, or 360-degree cameras). These detection systems are installed on strategic high points like towers for surveillance and quick detection, verification and tracking of the location of the fire (Barmpoutis et al., 2020). Camera-based monitoring systems are standard in the South African forestry landscape. Correctly located cameras scan and monitor the plantation areas and surrounding areas for smoke, and when smoke is detected, operators verify, validate and identify the fire location (Heyns et al., 2019). Even though autonomous fire detection and location systems are relatively accurate, the human observer is still an essential part of any detection system to validate and verify alerts and automatically

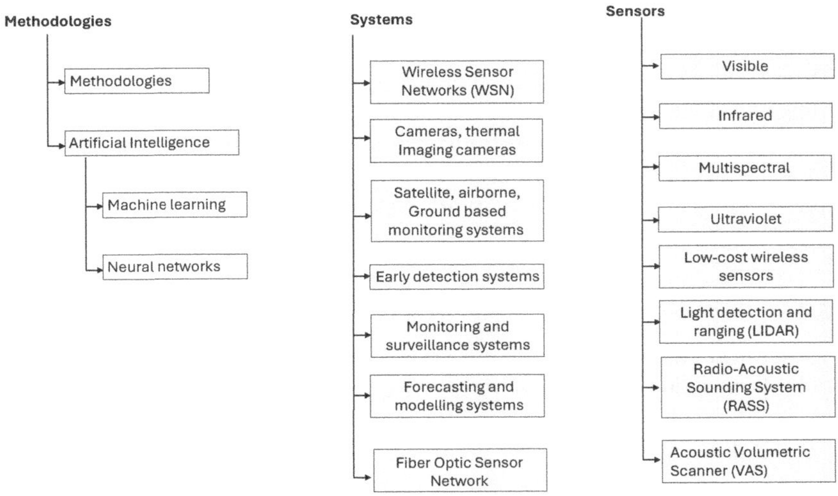

Figure 4.8 An overview of the main techniques for fire monitoring and detection (Carta et al., 2023)

derive fire-related information (Matthews et al., 2012). In the Brazilian context, almost 70% of the eucalyptus forest plantation surveyed land had some level of automatic fire monitoring. However, for pine plantations, just over 40% had the same level of technology for fire (Guerra et al., 2025).

On high fire danger days, the standby teams are positioned strategically across the plantations so that once a fire is detected, they can respond quickly. In South Africa, this approach is only applied during high fire danger periods. In Brazil, this approach represents almost 60% of the 221,555 hectares of pine and 30% of the 1,796,936 hectares of eucalyptus surveyed by Guerra et al. (2025).

Wireless sensors can sense their surroundings by detecting temperature, pressure, humidity, and chemical parameters (e.g. carbon dioxide) and computing the data (Alkhatib, 2014). These wireless sensors are not currently used in South Africa but have potential in future.

ii Aerial detection system

Aerial detection involves using an aeroplane to fly and monitor areas, detect fires, and inform fighting resources to extinguish them. This is not a common way of fire detection in South Africa or Brazil.

iii Satellite-based systems

Remote sensing by using satellites involves using imagery, aerial photography, and various sensing technologies to identify and record fire occurrences, burnt areas, and smoke clouds from burnt or burning areas (Carta et al., 2023). This detection system comprises earth-orbiting satellites that detect forest fires using a resolution imaging spectroradiometer. Satellites are helpful tools that provide essential data about fire danger rating prediction, fuel, fire mapping, fire monitoring, and general fire ecology information (Xiao-Rui et al., 2005). Satellites can be equipped with image sensors (e.g. advanced very-high-resolution radiometer, moderate resolution imaging spectroradiometer, visible infrared imaging radiometer), which provide specific spatial resolutions and can be used for fire detection; however, it is essential to note that earth mapping may take some time to complete (Carta et al., 2023).

4.5.2 Firefighting and suppression technologies

This section focuses on firefighting methods and associated technologies available to reduce fuel and suppress fires in the event of an occurrence.

i Manual methods

Manual firefighting methods are standard in South Africa and Brazil. This method is primarily effective in light fuels and uses trained manual labour with hand tools (e.g. fire beaters, rake hoes, and knapsack/backpack sprayers) to reduce fuel or extinguish an existing fire (Figure 4.9). It is slow and labour-intensive but flexible and can access areas other methods may struggle to access (Tadesse and Seboko, 2013).

ii Motor manual methods

Power tools such as blowers can be used for fuel clearing and fire extinguishing. The blowers use high-velocity air to blow light vegetation (e.g. leaves) away and create a clear firebreak instead of conventional hoes (Figure 4.10). Furthermore, blowers can consistently supply controlled air to extinguish flames instead of using traditional fire beaters (Ramantswana, 2022). Although blowers are practical tools, they should be used with other fire-extinguishing resources to ensure optimal fire suppression effectiveness. This method is not standard in Brazil, where less than 6% of

Figure 4.9 Manual firefighting

Figure 4.10 Motor manual forest protection using a blower

the eucalyptus surveyed area and 0% of the pine area were observed during the first firefighting index realised (Guerra et al., 2025).

iii Semi-mechanised methods

Semi-mechanised systems comprise a combination of machines with a complement of people to extinguish the fire. Fire vehicles act as the primary mover, carrying water and people to extinguish a fire. Fire vehicles of varying sizes with pumps can be used to extinguish fires effectively, as they provide a water source and personnel to apply water to the fire. Fire trucks that are purposefully designed for firefighting in forestry are designed to carry people, tools, water, and extendable pipes to enable people to combat fires effectively (Figure 4.11). Slip-on and drop-off units (bakkie sakkies) with capacities that range from 200 to 700 litres are versatile units that can be loaded on light-duty vehicles. Larger firefighting trucks can have capacities ranging from 1,500 to 5,500 and even more, depending on the design. Most forestry fire trucks are purpose-built for forestry conditions and comprise tanks mounted on the chassis, a crew cab, and a compartment area for storing pumps, tools and other equipment (Ramantswana, 2019). Trucks also have safety equipment such as sprinkler systems and remotely controlled cannons.

iv Mechanised methods

a Remotely piloted aircraft systems

The role of UAVs in monitoring, detecting, and fighting forest fires is becoming increasingly important. UAVs equipped with remote sensing vision systems are becoming a viable option for quick and cost-effective firefighting operations (Yuan et al., 2015).

b Aircraft systems (helicopter and fixed wing)

Helicopters and aeroplanes are vital tools for fire suppression, helping to control and limit the spread of fires in forests (Tadesse and Seboko, 2013). Aeroplanes (water bombers) (Figure 4.12) are generally faster and can carry more water than helicopters (Figure 4.13). However,

Figure 4.11 Firefighting truck

Figure 4.12 Fixed-wing aircraft for firefighting

helicopters are more versatile because they can refill water without needing to land, which improves their turnaround time. To maximise this method's effectiveness, ground crews should support aerial assets. Additionally, aerial resources need to be coordinated by a spotter to ensure that water is applied effectively and that safety protocols are followed during and after drops. Moreover, water efficiency in fire suppression can be enhanced by adding retardants and dampening agents. Aerial systems are quicker and can reach areas that ground crews might find difficult. Nevertheless, these systems are expensive, and their use is limited in extreme weather conditions, such as strong winds and poor visibility.

Figure 4.13 Helicopter used in firefighting

4.5.3 Integrated wildfire management

Integrated fire management is described as combining the various components of fire management (fire prevention, detection, suppression, and use) with consideration of ecological systems (Silva et al., 2010). It is an interconnected approach that includes fire awareness, prevention, prescribed burning, resource sharing, coordination, detection, fire suppression, and rehabilitation (FynbosFire, 2016). It is a holistic method that applies fire in diverse ways to ensure that natural ecosystems and manmade infrastructure resources are appropriately managed, thereby limiting negative impacts. Suppression alone is not sufficient to deal with forest fires; there must be an appreciation of the ecological, social, cultural, and economic roles in which fires occur, and an integrated response should be developed to manage fires effectively. According to Moore (2019), integrated fire management has evolved over time. It fundamentally involves people (activities and capabilities of various stakeholders), fire in the landscape (considering fire holistically), and activities (fire management systems). The effectiveness of integrated fire management is greatly improved by using fire detection, fire behaviour monitoring, and forecasting technologies to collect data and enhance fire control strategies.

4.6 Thinning

4.6.1 Overview of thinning

Thinning as a silvicultural practice can influence stand structure, tree and stand growth, products, yields, and diversity (Gonçalves, 2021). Thinning is an intervention carried out to decrease the number of stems at the end of the rotation and overall competition within a stand, allowing the remaining trees to grow faster and wider in diameter (del Campo et al., 2022). It also helps utilise trees that might otherwise be lost due to natural mortality, pests, or diseases (David et al., 2017). Thinning can be performed alongside pruning as a key silvicultural practice that influences stand structure, growth, products, yields, and diversity (Gonçalves, 2021), and it can improve wood quality (Lopes et al., 2013, de Oliveira et al., 2012). If all other conditions remain constant (e.g. nutrients and water), the remaining trees have a growth advantage because they are well-spaced and will grow more quickly than densely planted trees (Malan and Hoon, 1992).

Several thinning methods are available depending on landowners' goals and the desired future characteristics of the trees, with standard methods including thinning from below, thinning from above, and selective thinning (Gonçalves, 2021). In South Africa, the 7th row thinning is typically the initial thinning carried out on a sawtimber stand (Brink, 2012). Afterwards, marked trees, such as small, deformed, and unsuitable ones, are primarily removed, and additional trees are eliminated to achieve the targeted tree distribution and stand density (see Table 4.4).

According to Kerr and Haufe (2011), the main objectives of thinning are:

- Present conditions for a healthy growing stand
- Eliminate trees with undesirable characteristics (e.g. form and size)
- Provide enough space for remaining trees to grow and produce sawlog timber with large diameters
- Maximise economic returns during the stand rotation
- Maintain canopy closure for environmental preservation of the soil

Thinning can affect stand development differently, depending on the intensity and timing (age) at which the trees were removed. Thinning can be conducted using various harvesting methods, i.e. cut-to-length, tree-length and full-tree

Table 4.4 Examples of thinning regimes for pine sawtimber in South Africa

Planting density and spacing						
Trees/ha Spacing	1,372 trees/ha 2.7 × 2.7 m		1,111 Trees/ha 3.0 × 3.0 m		816 Trees/ha 3.5 × 3.5 m	
	Age (years)	*Trees/ha*	*Age (years)*	*Trees/ha*	*Age (years)*	*Trees/ha*
Establishment	0	1,372	0	1,111	0	816
1st thinning	8–10	650–750	8–10	450–750	9–13	450–500
2nd thinning	11–15	400–500	12–15*	300–500	13–18*	275–300
3rd thinning	14–18*	275–300	17*	300		
Clear-fell	24–30	0	24–30	0	24–30	0

Adapted from Brink (2012).

Thinnings marked with (*) are mostly optional for good site qualities.

method (Langin et al., 2010). The technology levels associated with the harvesting method can be manual, semi-mechanised, or fully mechanised. There are three types of thinnings practised in forestry operations:

- Systematic thinning – trees removed according to a specific system without considering characteristics of specific trees, e.g. row thinning or strip thinning (Kerr and Haufe, 2011)
- Selective thinning – trees are removed (e.g. competing trees, trees with defects, poor form and diseased trees) or kept (good form and size) based on specific characteristics they possess (Kerr and Haufe, 2011)
- Combination thinning – is the silvicultural operation that combines two thinning methods – systematic and selective thinning, offering speed and efficiency (Makoto, 2021, Fritscher, 2022).

Compared to clear-felling, thinning costs are higher, and productivity is lower (Ackerman et al., 2016). The productivity of thinning machines can be affected by factors such as tree size, marking of trees, number of timber assortments, time of day operation is conducted (day or night), terrain conditions, operator skills and travel distance (Pētersons, 2010, Sirén and Aaltio, 2003, Chang et al., 2023, Lageson, 1997).

4.6.2 Semi-mechanised thinning

Semi-mechanised thinning is conducted using a combination of manual and mechanised methods to fell, process, and extract the timber to the roadside or depot. Thinning technology must enable the selection of desired trees while causing minimal damage to the residual trees. Thinning can be conducted using the cut-to-length (CTL), tree-length (TL), or full-tree (FT) harvesting method. The semi-mechanised harvesting systems comprise motor manual felling combined with mechanised extraction using tractors, including those with an A-frame, tractor trailers, tractors equipped with single and double drum winch and cable yarding systems (Han and Han, 2020). Dembure et al. (2019) found that when conducting a second commercial thinning in *Pinus elliottii* stands, the CTL system (harvester and forwarder) resulted in higher productivity and less residual stand damage compared to the TL system (chainsaw for felling, delimbing, and topping, three-wheeled loggers, winch-fitted tractor, and chainsaw for roadside processing). The use of whole tree harvesting using chainsaw felling and single drum tractor extraction is productive and cost-effective in moderate to heavy thinning operations (Proto et al., 2020).

4.6.3 Mechanised thinning

Mechanised thinning predominantly uses machinery like harvesters and forwarders to perform the cutting and extraction operations for the extensive

Pinus plantations in Brazil, not in eucalypt plantations (Weber et al., 2013, da Silva Lopes et al., 2016, da Silva Lopes et al., 2017). It could be a complex and economically demanding operation due to limited space within the forest stand, low machinery productivity, elevated production expenses, and lower market value of the thinned trees (da Silva Lopes et al., 2016). This silvicultural operation, removing biomass from the stand, impacts site nutrient levels, so leaving bark and other harvesting residues, such as needles, branches, and tops, on the forest floor can contribute to nutrient cycling (Viera et al., 2011).

Mechanised thinning operations are conducted using machines to fell, process, and extract trees to the roadside or depot. Mechanised CTL equipment comprises a harvester (tracked or wheeled) and a forwarder (Makoto, 2021, Sirén and Aaltio, 2003). A full tree can comprise a feller buncher and skidder (grapple or clambunk) with a roadside processor. In thinning operations, whole tree (feller buncher and skidders) are generally more productive and cost-effective than CTL and TL harvesting systems (Chang et al., 2023). However, whole tree harvesting removes the tree crown (tops and leaves), which serves as a source of nutrients for the site. Therefore, it is essential to consider residual stand damage and the impact on the site productivity over and above the harvesting machines' productivity and costs (Jacobson et al., 2000, Nord-Larsen, 2002)

4.7 Summary

This chapter focused on silviculture technologies in stand tending, including weeding after planting, forest protection, pruning, and thinning. It detailed how weeding after planting is carried out using manual, motor manual, mechanical, and chemical methods, highlighting the prevalence of chemical control in South Africa and Brazil. This chapter discussed forest protection from pests, diseases, and wildfires, noting integrated pest management strategies and various detection and suppression technologies. Pruning was examined as a wood quality improvement operation, using both manual and motor manual techniques. Lastly, thinning was described as a practice to influence stand structure and growth, detailing systematic, selective, and combination thinning methods, along with semi-mechanised and mechanised approaches.

Review questions

1. Differentiate between chemical and mechanical weed control after planting by referring to various technologies used to perform both operations
2. Differentiate between manual and motor manual pruning
3. Name and describe technologies that can be used for forest protection against various pests and diseases
4. What is integrated fire management, and how can it be improved by using advanced technologies

5 Describe different harvesting methods and systems that can be used to perform mechanised thinning in forest stands

References

Ackerman, S., Fryer, D. & Jackson, R. 2018. *South African Outdoor Power Equipment Safety & Operating Handbook*. Pietermaritzburg: Institute for Commercial Forestry Research.

Ackerman, S. A., Seifert, S., Ackerman, P. A. & Seifert, T. 2016. Mechanised pine thinning harvesting simulation: Productivity and cost improvements as a result of changes in planting geometry. *Croatian Journal of Forest Engineering: Journal for Theory and Application of Forestry Engineering,* 37, 1–12.

Alkhatib, A. A. 2014. A review on forest fire detection techniques. *International Journal of Distributed Sensor Networks,* 10, 597368.

Barbaresco, V., Guerra, S. &Oguri, O. 2021. *Avaliação comparativa entre implementos no preparo de solo com escavadora*. Piracicaba: IPEF.

Barmpoutis, P., Papaioannou, P., Dimitropoulos, K. & Grammalidis, N. 2020. A review on early forest fire detection systems using optical remote sensing. *Sensors,* 20, 6442.

Botha, M., Kotze, H. & DU Toit, B. 2012. Silviculture of industrial Eucalyptus and Acacia plantations in South Africa. *In:* Bredenkamp, B. & Upfold, S. (eds.) *South African Forestry handbook.* Pretoria: South African Institute of Forestry Research.

Brasil. 2021. Estabelece regras para operação de aeronaves remotamente pilotadas destinadas à aplicação de agrotóxicos e afins, adjuvantes, fertilizantes, inoculantes, corretivos e sementes. *In:* Ministério da Agricultura, Pecuária e Abastecimento (ed.) *Portaria MAPA nº 298, de 22 de setembro de 2021*. Brasília, DF: Diário Oficial da União, Seção 1, 14.

Brink, M. 2012. Certification as a risk management strategy. *In:* Bredenkamp, B. & Upfold, S. (eds.) *South African Forestry handbook.* Pretoria: South African Institute of Forestry Research.

Brockerhoff, E. G., Liebhold, A. M., Richardson, B. & Suckling, D. M. 2010. Eradication of invasive forest insects: Concepts, methods, costs and benefits. *New Zealand Journal of Forestry Science,* 40, S117–S135.

Byrnes, W., Krajicek, J. & Wichman, J. 1973. Weed control. Black walnut as a crop, Black Walnut Symposium, USDA Forest Service, North Central Forest Experiment Station, St. Paul, MN, Gen. Tech. Rep. NC–4, 42–48.

Carta, F., Zidda, C., Putzu, M., Loru, D., Anedda, M. & Giusto, D. 2023. Advancements in forest fire prevention: A comprehensive survey. *Sensors,* 23, 6635.

Carvalho, F. K., Chechetto, R. G., Mota, A. A. & Antuniassi, U. R. 2020. Challenges of aircraft and drone spray applications. *Outlooks on Pest Management,* 31, 83–88.

Chang, H., Han, H.-S., Anderson, N., Kim, Y.-S. & Han, S.-K. 2023. The cost of Forest thinning operations in the Western United States: A systematic literature review and new thinning cost model. *Journal of Forestry,* 121, 193–206.

Chen, G. & Meentemeyer, R. K. 2016. Remote sensing of forest damage by diseases and insects. *In:* Weng, Q. (ed.) *Remote Sensing for Sustainability*. CRC Press, Boca Raton 145–162.

Chungu, D., Muimba-Kankolongo, A., Wingfield, M. J. & Roux, J. 2010. Plantation forestry diseases in Zambia: Contributing factors and management options. *Annals of Forest Science,* 67, 802.

Cornago Junior, V. M. 2020. *Caracterização da capacidade operacional de pulverizadores de barras no Brasil dos anos de 2008 a 2018*. Master dissertation, Universidade Estadual Paulista, Faculdade de Ciências Agronômicas, Botucatu.

da Silva Cavalcante, W. S., DA Silva, N. F., Teixeira, M. B., Neto, G. Z., Cabral Filho, F. R., Cunha, F. N. & Corrêa, F. R. 2022. Tecnologias e inovações no uso de drones na agricultura Technologies and innovations in the use of drones in agriculture. *Brazilian Journal of Development,* 8, 7108–7117.

Da Silva Lopes, E., Diniz, C. C. C., Serpe, E. L. & Cabral, O. M. D. J. V. 2016. Efeito do sortimento da madeira na produtividade e custo do forwarder no desbaste comercial de Pinus taeda. *Scientia Forestalis,* 44, 57–66.

Da Silva Lopes, E., Roza, B. L. & De Oliveira, F. M. 2017. Efeito de variáveis operacionais na produtividade de um harvester de pneus no desbaste de pinus. *Floresta,* 47.

David, H. C., Péllico, S., Arce, J. E., Corte, A. P. D., Marinheski Filho, A. & Araújo, E. J. G. D. 2017. Efeito da qualidade do sítio e do desbaste na produção de pinus. *Floresta e Ambiente,* 24, e00096414.

De Oliveira, F. M., Da Silva Lopes, E., Malinovski, J. R., Da Silva, R. H. & Rodrigues, C. K. 2012. Avaliação técnica e de custos de poda manual e semimecanizada em plantios de Pinus taeda. *Floresta,* 42, 691–700.

Del Campo, A. D., Otsuki, K., Serengil, Y., Blanco, J. A., Yousefpour, R. & Wei, X. 2022. A global synthesis on the effects of thinning on hydrological processes: Implications for forest management. *Forest Ecology and Management,* 519, 120324.

Dembure, T. P., Mcewan, A., Spinelli, R., Magagnotti, N. & Ramantswana, M. 2019. A comparison between two alternative harvesting systems in the thinning of fast-growing pine plantations under the conditions of low labour cost. *European Journal of Forest Research,* 138, 43–52.

Den Breejen, E., Breuers, M., Cremer, F., Kemp, R., Roos, M., Schutte, K. & De Vries, J. S. 1998. Autonomous forest fire detection. *In:* Viegas, D. X. (ed.) *Proceedings of the III International Conference on Forest Fire Research and 14th Conference on Fire and Forest Meteorology*. Vol. II. Coimbra: ADAI-Associacao para o Desenvolvimento da Aerodinamica Industrial, 2003–2012.

Deng, Y., Yang, G., Xie, Z., Yu, J., Jiang, D. & Huang, Z. 2020. Effects of different weeding methods on the biomass of vegetation and soil evaporation in eucalyptus plantations. *Sustainability,* 12, 3669.

Donald, D. 1986. Cleaning operations in South African forestry. *South African Forestry Journal,* 137, 38–49.

Dornelles, M. E., Schlosser, J. F., Boller, W., Russini, A. & Casali, A. L. 2011. Inspeção técnica de tratores e pulverizadores utilizados em pulverização agrícola. *Revista Engenharia na Agricultura-Reveng,* 19, 36–43.

Ehrentraut, G. & Branter, K. 1990. Vegetation management by manual and mechanical means in Alberta boreal forests. *The Forestry Chronicle,* 66, 366–368.

Eugenio, F. C., Schons, C. T., Mallmann, C. L., Schuh, M. S., Fernandes, P. & Badin, T. L. 2020. Remotely piloted aircraft systems and forests: a global state of the art and future challenges. *Canadian Journal of Forest Research,* 50, 705–716.

Evans, J. & Turnbull, J. W. 2004. *Plantation Forestry in the Tropics: The Role, Silviculture, and Use of Planted Forests for Industrial, Social, Environmental, and Agroforestry Purposes*. Oxford: Oxford University Press.

Ferreira, J. C., Stähelin, T. S. F., Jesus, M. S. D., Muñiz, G. I. B. D., Brand, M. A. & Freitas, T. P. 2019. Estimativa da oferta de biomassa florestal em povoamentos de Pinus taeda L. após intervenções culturais. *Ciência Florestal,* 29, 1459–1468.

Forest Stewardship Council. 2019. *SC Pesticides Policy*. Bonn: FSC.

Fritscher, J. L. 2022. Manejo no cultivo de Pinus spp. *Trabalho de conclusão de curso,* Universidade Federal do Rio Grande do Sul, Faculdade de Agronomia.

Fynbosfire. 2016. *The Integrated Fire Management Handbook*. Cape Town: Western Cape.

Gadgil, P. D. & Bain, J. 1999. Vulnerability of planted forests to biotic and abiotic disturbances. *New Forests,* 17, 227–238.

Godsmark, R. & Oberholzer, F. 2019. The South African forestry and forest products industry 2017. *Annual Stats of SA Forestry Industry*. Pietermaritzburg: Forestry South Africa, 1–160.

Goldsmith, A. A. 2019. *The Combined Effect of Slash Burning and Repeated Disc Harrowing on Changes in Fuel Loading, Soil Properties, Root Growth and Stand Productivity of Eucalypts in Mpumalanga: South Africa*. Stellenbosch: Stellenbosch University.

Gonçalves, A. C. 2021. *Thinning: An Overview*. London: IntechOpen, 41–58.

Gonçalves, J. D. M., Stape, J. L., Laclau, J.-P., Bouillet, J.-P. & Ranger, J. 2008. Assessing the effects of early silvicultural management on long-term site productivity of fast-growing eucalypt plantations: The Brazilian experience. *Southern Forests: A Journal of Forest Science,* 70, 105–118.

Guerra, S. P. S., Oguri, G., Schumacher, M. V., Witschoreck, R. & Balista, B. Z. 2025. *Survey of the Level of Mechanization in Forestry: 2024/2025 Edition*. Piracicaba: IPEF.

Hafeez, A., Husain, M. A., Singh, S., Chauhan, A., Khan, M. T., Kumar, N., Chauhan, A. & Soni, S. 2023. Implementation of drone technology for farm monitoring & pesticide spraying: A review. *Information Processing in Agriculture,* 10, 192–203.

Han, S.-K. & Han, H.-S. 2020. Productivity and cost of whole-tree and tree-length harvesting in fuel reduction thinning treatments using cable yarding systems. *Forest Science and Technology,* 16, 41–48.

Herdiyeni, Y., Jamaluddin, M. I., Setio, T., Dewanto, V., Tjahjono, B. & Siregar, B. A. 2017. An integrated smart surveillance system for diseases monitoring in tropical plantation forests. *2017 IEEE 17th International Conference on Communication Technology (ICCT),* Chengdu, China: IEEE, 1822–1826.

Heyns, A., Du Plessis, W., Kosch, M. & Hough, G. 2019. Optimisation of tower site locations for camera-based wildfire detection systems. *International Journal of Wildland Fire,* 28, 651–665.

Hoppe, J. M. & Freddo, Á. R. 2003. Efeito da intensidade de desrama na produção de Pinus elliottii Engelm., no município de Piratini, RS. *Ciência Florestal,* 13, 47–56.

Jacobson, S., Kukkola, M., Mälkönen, E. & Tveite, B. 2000. Impact of whole-tree harvesting and compensatory fertilization on growth of coniferous thinning stands. *Forest Ecology and Management,* 129, 41–51.

Jones, F. 2019. *Pulverização por drones* [Online]. Revista Pesquisa Fapesp. Available: https://revistapesquisa.fapesp.br/pulverizacao-por-drones [Accessed 27 December 2024].

Jordan, C., Dos Santos, P. L., Oliveira, L. R. D. S., Domingues, M. M., Gêa, B. C. C., Ribeiro, M. F., Mascarin, G. M. & Wilcken, C. F. 2021. Entomopathogenic fungi as the microbial frontline against the alien Eucalyptus pest Gonipterus platensis in Brazil. *Scientific Reports,* 11, 7233.

Kerr, G. & Haufe, J. 2011. Thinning practice: A silvicultural guide. *Forestry Commission,* 1, 5–54.

Kraaij, T., Baard, J. A., Arndt, J., Vhengani, L. & Van Wilgen, B. W. 2018. An assessment of climate, weather, and fuel factors influencing a large, destructive wildfire in the Knysna region, South Africa. *Fire Ecology,* 14, 1–12.

Lageson, H. 1997. Effects of thinning type on the harvester productivity and on the residual stand. *Journal of Forest Engineering,* 8, 7–14.

Langin, D., Ackerman, P., Krieg, B., Immelman, A., Potgieter, C., van Rooyen, J. & Upfold, S. 2010. *South African Ground Based Harvesting Handbook.* Scottsville: Forest Engineering Southern Africa and Institute for Commercial Forestry Research (ICFR). ISBN 978-0-620-49065-8.

Lantagne, D. O. & Burger, J. A. 1987. Comparison of site preparation methods for weed control in loblolly pine (Pinus taeda) plantations. *Weed Science,* 35, 590–593.

Lopes, E. D. S., Oliveira, F. M. D. & Rodrigues, C. K. 2011. Determinação da carga física de trabalho na atividade de aplicação manual de herbicida Determination of demanded physical effort in herbicide application activity. *Ambiência,* 7, 329–337.

Lopes, E. S., Oliveira, F. M., Malinovski, J. R. & DA Silva, R. H. 2013. Avaliação biomecânica de trabalhadores nas atividades de poda manual e semimecanizada de pinus taeda. Floresta, Curitiba, PR, 43, 9–18.

Machado-Neto, J., Bassini, A. & Aguiar, L. 2000. Safety of working conditions of glyphosate applicators on Eucalyptus forests using knapsack and tractor powered sprayers. *Bulletin of Environmental Contamination & Toxicology,* 64.

Madeira, M., Fabiao, A. & Carneiro, M. 2012. Do harrowing and fertilisation at middle rotation improve tree growth and site quality in Eucalyptus globulus Labill. plantations in Mediterranean conditions? *European Journal of Forest Research,* 131, 583–596.

Makoto, M. 2021. *Evaluation of a Mechanized CTL Harvesting System in Different Geometric Thinning Operations in Terms of Productivity, Cost, Biomass Conversion, Residual Stand Damage and Spacing Uniformity.* Stellenbosch: Stellenbosch University.

Malan, F. & Hoon, M. 1992. Effect of initial spacing and thinning on some wood properties of Eucalyptus grandis. *South African Forestry Journal,* 163, 13–20.

Mallik, A. U., Bell, F. W. & Gong, Y. 2002. Effectiveness of delayed brush cutting and herbicide treatments for vegetation control in a seven-year-old jack pine plantation in northwestern Ontario, Canada. *Silva Fennica,* 36, 505–519.

Matthews, S., Sullivan, A., Gould, J., Hurley, R., Ellis, P. & Larmour, J. 2012. Field evaluation of two image-based wildland fire detection systems. *Fire Safety Journal,* 47, 54–61.

Miller, J. H. & Mitchell, R. J. 1990. *A Manual on Ground Applications of Forestry Herbicides*. Atlanta, GA: USDA Forest Service, Southern Region.

Montagu, K., Kearney, D. & Smith, R. 2003. The biology and silviculture of pruning planted eucalypts for clear wood production—A review. *Forest Ecology and Management,* 179, 1–13.

Monteiro, A. & Santos, S. 2022. Sustainable approach to weed management: The role of precision weed management. *Agronomy,* 12, 118.

Moore, P. F. 2019. Global wildland fire management research needs. *Current Forestry Reports,* 5, 210–225.

Morris, A. R. 2022. Changing use of species and hybrids in South African forest plantations. *Southern Forests: A Journal of Forest Science,* 84, 193–205.

Naskrent, B., Grzywiński, W., Łukowski, A. & Polowy, K. 2020. Influence of cutting attachment on noise level emitted by brush cutter during tending of young forests.

Croatian Journal of Forest Engineering: Journal for Theory and Application of Forestry Engineering, 41, 129–135.

Ndlovu, N. N., Little, K., Baillie, B. & Rolando, C. 2022. An evaluation of the environmental behaviour, fate and risk of key pesticides used in South African forest plantations. *Southern Forests: A Journal of Forest Science,* 84, 83–92.

Nord-Larsen, T. 2002. Stand and site productivity response following whole-tree harvesting in early thinnings of Norway spruce (Picea abies (L.) Karst.). *Biomass and Bioenergy,* 23, 1–12.

Oregon State University. 2024. *Weeds* [Online]. Online: Oregon State University. Available: https://forages.oregonstate.edu/nfgc/eo/onlineforagecurriculum/instructormaterials/availabletopics/weeds/herbicides [Accessed 2 May 2024].

Park, B. B., Seo, J. M., Han, S. H., Youn, W. B., Jung, Y. K., Namgung, B. S., Lee, S. J. & Lee, S. I. 2020. Analysis of the current status of weeding operation and crop tree growth across planting periods. *Journal of Korean Society of Forest Science,* 109, 179–188.

Payn, T., Carnus, J.-M., Freer-Smith, P., Kimberley, M., Kollert, W., Liu, S., Orazio, C., Rodriguez, L., Silva, L. N. & Wingfield, M. J. 2015. Changes in planted forests and future global implications. *Forest Ecology and Management,* 352, 57–67.

Pereira, D. P., Fiedler, N. C., Lima, J. S. D. S., Guimarães, P. P., Môra, R. & Carmo, F. C. D. A. D. 2012. Eficiência da subsolagem na profundidade de preparo do solo em função da declividade do terreno. *Cerne,* 18, 607–612.

Peruzzi, A., Martelloni, L., Frasconi, C., Fontanelli, M., Pirchio, M. & Raffaelli, M. 2017. Machines for non-chemical intra-row weed control in narrow and wide-row crops: a review. *Journal of Agricultural Engineering,* 48, 57–70.

Pētersons, J. Factors affecting harvester productivity in forest thinning in Latvia. With this issue of 2010, we bring 83 proceedings of the 108, which started life as presentations at the Annual 16th International Scientific Conference "Research for Rural Development 2010" held at the Latvia University of Agriculture, in Jelgava, on 19 to 21 May 2010. In the retrospect of four months later, we can count the Conference as a great success. The theme–Research for Rural Development-attracted participation of 292 researchers with very different, 2010. 183.

Pinkard, E. & Beadle, C. 2000. A physiological approach to pruning. *The International Forestry Review,* 2, 295–305.

Programa Cooperativo Sobre Mecanização E Automação Florestal. 2020. *13a. Reunião de Filiadas do PCMAF*. Piracicaba: IPEF.

Programa Cooperativo Sobre Mecanização E Automação Florestal. 2022. *14a Reunião de Filiadas do PCMAF*. Três Lagoas: IPEF.

Programa Cooperativo Sobre Mecanização E Automação Florestal. 2025. *Relatório de visita técnica*. Ribeirão Preto: IPEF.

Proto, A. R., Bernardini, V., Cataldo, M. F. & Zimbalatti, G. 2020. Whole tree system evaluation of thinning a pine plantation in southern Italy. *Annals of Silvicultural Research,* 45, 44–52.

Rajendran, P., Parthiban, K., Nandhagopalan, A., Devanand, P. & Kumar, P. 2014. 20 Mechanization in forestry operations. *In:* Parthiban, K. T., Umarani, R., Umesh Kanna, S., Sekar, I., Rajendran, P. & Durairasu, P. (eds.) *Industrial Agroforestry Perspectives and Prospectives*. Jodhpur: Scientific Publishers, 303–319. ISBN 9788172339050.

Ramantswana, M., Brink, M., Little, K., Spinelli, R. & Chirwa, P. 2021. Perspectives and drivers of modernisation of silviculture re-establishment in South Africa. *Southern Forests: A Journal of Forest Science,* 83, 79–87.

Ramantswana, M. M., Brink, M. P., Little, K. M., Spinelli, R. & Chirwa, P. W. 2020. Current status of technology-use for plantation re-establishment in South Africa. *Southern Forests: A Journal of Forest Science,* 82, 313–323.

Ramantswana, N. 2019. *Engineering Principles*. George: Nelson Mandela University.

Ramantswana, N. 2022. *STIHL's BR 800 C-E Backpack Blower for Fire Management* [Online]. Forsilvitech website: Forsilvitech. Available: https://www.forsilvitech.com/article/stihl-s-br-800-c-e-backpack-blower-for-fire-management/ [Accessed 24 February 2025 2025].

Roberts, J., Little, K. & Rolando, C. 2021. Estimated herbicide use in the commercial forest sector in South Africa. *Australian Forestry,* 84, 108–121.

Robinne, F.-N. 2021. Impacts of disasters on forests, in particular forest fires. *UNFFS Background paper*.

Roux, J., Hurley, B. & Wingfield, M. 2012. Diseases and pests of eucalypts, pines and wattle. *In:* Bredenkamp, B. & Upfold, S. (eds.) *South African Forestry Handbook*. Pretoria: South African Institute of Forestry Research.

Rullan-Silva, C. D., Olthoff, A. E., Delgado De la Mata, J. A. & Pajares-Alonso, J. A. 2013. Remote monitoring of forest insect defoliation. A review. *Forest Systems,* 22 (3), 377–391. https://doi.org/10.5424/fs/2013223-04417.

Salomão, P. E. A., Ferro, A. M. S. & Ruas, W. F. 2020. Herbicidas no Brasil: um breve revisão. *Research, Society and Development,* 9, e32921990–e32921990.

Santos, K. F. D. & Reichert, J. M. 2022. Best tillage practices for eucalyptus growth and productivity: A review on the Brazilian experience. *Revista Brasileira de Ciência do Solo,* 46, e0210091.

Scolforo, J. R. S., Acerbi, F. W., Oliveira, A. D. D. & Maestri, R. 2001. Simulação e avaliação econômica de regimes de desbastes e desrama para obter madeira de Pinus taeda livre de nós. *Ciência florestal,* 11, 121–139.

Shekwa, A., Chirwa, P., Ngulube, E. & Ghebremariam, T. 2017. Productivity, efficiency and costs of manual saw and electric shear pruning operations in Pinus elliottii stands of Mpumalanga, South Africa. *International Journal of Forest Engineering,* 28, 169–175.

Shikwambana, L., Ncipha, X., Malahlela, O. E., Mbatha, N. & Sivakumar, V. 2019. Characterisation of aerosol constituents from wildfires using satellites and model data: A case study in Knysna, South Africa. *International Journal of Remote Sensing,* 40, 4743–4761.

Sibiya, Z. B., Ackerman, P. & Ackerman, S. A. 2021. Productivity and workload analysis between manual and motor-manual pruning of Pinus patula at two different lifts. *International Journal of Forest Engineering,* 32, 191–201.

Silva, J. R. V. D., Alves, P. L. D. C. A. & Toledo, R. E. B. D. 2012. Weed control strip influences the initial growth of Eucalyptus grandis. *Acta Scientiarum. Agronomy,* 34, 29–35.

Silva, J. S., Rego, F., Fernandes, P. & Rigolot, E. 2010. *Towards Integrated Fire Management: Outcomes of the European Project Fire Paradox*. Joensuu: European Forest Institute Joensuu.

Sirén, M. & Aaltio, H. 2003. Productivity and costs of thinning harvesters and harvester-forwarders. *International Journal of Forest Engineering,* 14, 39–48.

Stone, C. & Mohammed, C. 2017. Application of remote sensing technologies for assessing planted forests damaged by insect pests and fungal pathogens: A review. *Current Forestry Reports,* 3, 75–92.

Strydom, S. & Savage, M. J. 2016. A spatio-temporal analysis of fires in South Africa. *South African Journal of Science,* 112, 1–8.

Swift, R. 2020. *Flail Mower vs Rotary Mower – Which Is Best for Your Job?* [Online]. Online: Diamond Mowers. Available: https://blog.diamondmowers.com/flail-mower-vs-rotary-mower [Accessed 06 May 2024 2024].

Tadesse, E. & Seboko, B. 2013. Forest *Wildland Fire Prevention and Control for Sustainable Forest Management.* Wondo Genet: Hawassa Uniiversiity.

Teka, A., Yirefu, F. & Abadi, Z. 2012. Increasing the efficiency of knapsack sprayers by modifying a single nozzle sprayer into a low cost multi–nozzle sprayer. *Ethiopian Journal of Health Development,* 5, 28–42.

TIPWG. 2024. *Approved Pesticide List* [Online]. Forestry South Africa. Available: https://www.tipwg.co.za/ [Accessed 2 May 2024].

Vasic, V., Konstantinovic, B. & Orlovic, S. 2012. Weeds in forestry and possibilities of their control. *Weed Control.* IntechOpen.

Viera, M., Schumacher, M. V. & Bonacina, D. M. 2011. Biomassa e nutrientes removidos no primeiro desbaste de um povoamento de Pinus taeda L. em Cambará do Sul, RS. *Revista Árvore,* 35, 371–379.

Vuaden Junior, N. J. 2019. *Capacitações e treinamentos em tecnologia de aplicação e rendimento operacional em pulverização.* Trabalho de conclusão de curso, Universidade Federal do Rio Grande do Sul, Faculdade de Agronomia, Porto Alegre, RS. Available: https://lume.ufrgs.br/handle/10183/234085.

Warkotsch, W. 1988. Harvesting of pine and eucalypt in South Africa. *South African Forestry Journal,* 147, 37–52.

Weber, V. P., Finger, C. A. G., Fleig, F. D., Souza, C. A. M. D., Munareto, F. F. & Silva, L. T. D. M. D. 2013. Diretrizes para desbaste de Pinus taeda L. em função da altura dominante. *Ciência Florestal,* 23, 193–201.

Wingfield, M., Roux, J., Coutinho, T., Govender, P. & Wingfield, B. 2001. Plantation disease and pest management in the next century. *Southern African Forestry Journal,* 2001, 67–72.

Wingfield, M., Slippers, B., Hurley, B., Coutinho, T., Wingfield, B. & Roux, J. 2008. Eucalypt pests and diseases: growing threats to plantation productivity. *Southern Forests: A Journal of Forest Science,* 70, 139–144.

Wingfield, M. J. 1990. Current status and future prospects of forest pathology in South Africa. *South African Journal of Science,* 86, 60–62.

Xiao-Rui, T., Mcrae, D. J., LI- Fu, S., Ming- Yu, W. & Hong, L. 2005. Satellite remote-sensing technologies used in forest fire management. *Journal of Forestry Research,* 16, 73–78.

Yuan, C., Zhang, Y. & Liu, Z. 2015. A survey on technologies for automatic forest fire monitoring, detection, and fighting using unmanned aerial vehicles and remote sensing techniques. *Canadian Journal of Forest Research,* 45, 783–792.

Zhang, J., Huang, Y., Pu, R., Gonzalez- Moreno, P., Yuan, L., Wu, K. & Huang, W. 2019. Monitoring plant diseases and pests through remote sensing technology: A review. *Computers and Electronics in Agriculture,* 165, 104943.

Appendix 1

Approved chemicals used in the South African forestry industry (TIPWG, 2024)

Herbicide	*Uses*
Clethodim	For pine and eucalypt compartments and industrial areas only (apart from Series 240). An emulsifiable concentrate herbicide for the post-emergence control of annual grasses
Clopyralid pyridine compound	For the control of *A. mearnsii* and other weeds listed on the label. Only in eucalypts and pine
Clopyralid monoethanolamine salt	For the control of *A. mearnsii* and other weeds listed on the label. Only in eucalypts and pine
Clopyralid triethylamine salt/ Triclopyr triethylamine salt	A systemic water-soluble concentrate herbicide for the control of certain undesirable woody plants in forestry, pastures, conservation, and non-crop areas. Confront and Contest can be used in industrial areas too
Cycloxydim	A selective, emulsifiable systemic weed killer for the post-emergence control of certain grasses in broadleaf and forestry crops
Flumioxazin	A herbicide for pre-emergence control of annual broadleaf weeds
Fluroxypyr meptyl ester/ Picloram	A systemic micro-emulsion herbicide for the control of woody plants and weeds
Fluroxypyr meptyl ester	An emulsifiable concentrate selective herbicide for the control of woody weeds
Fluroxypyr meptyl ester + Triclopyr butoxy ethyl ester	A systemic emulsifiable concentrate herbicide for the control of problem plants
Glufosinate ammonium	A non-selective, non-residual, contact herbicide in a water-soluble concentrate formulation, for tracer belt preparation
Glyphosate (glycine) (glyphosate ammonium salt)	A soluble concentrate non-selective systemic post-emergence herbicide with slight or no soil activity for the control of perennial and annual weeds
Glyphosate (glycine) (glyphosate isopropylamine salt)	

(Continued)

Herbicide	*Uses*
Glyphosate isopropylamine salt	A systemic herbicide for the control of weeds and grasses
Glyphosate (glycine) (glyphosate potassium salt)	A soluble concentrate non-selective systemic post-emergence herbicide with slight or no soil activity for the control of perennial and annual weeds. Only Roundup Turbo registered for conservation areas
Glyphosate ammonium salt	
Glyphosate N-(phosphonomethyl) glycine	
Haloxyfop-P-methyl ester	An emulsifiable concentrate selective, systemic post-emergence herbicide for the control of annual grasses
Imazapyr	A water-soluble non-selective herbicide for the control of perennial and annual weed plants
Metazachlor	For pine and eucalypt compartments. A suspension concentrate herbicide for the pre-emergence and early post-emergence control of annual grasses and certain broadleaf weeds. Yellow nutsedge may also be controlled under favourable conditions
Metsulfuron-methyl	A water-dispersible granular herbicide for the control of certain broad-leaved weeds as well as invader weeds. Note: only Forester is registered for use in compartments
Picloram potassium salt	
Picloram potassium salt/ Triclopyr trimethylamine salt	A ready-to-use water-based gel, paint-on systemic herbicide acting through the cut surfaces of certain woody plants
Picloram potassium salt/ Fluroxpyr	A ready-to-use water-based gel, paint-on systemic herbicide acting through cut surfaces of certain woody plants
Propaquizafop	An emulsifiable concentrate herbicide for the selective post-emergence control of annual grasses in crops, as indicated on the label
S-metolachlor	Pre-emergence herbicide for use in all compartments
Triclopyr butoxy ethyl ester	A systemic emulsifiable concentrate herbicide for the control of woody plants and weeds
Triclopyr triethylamine salt	A water-soluble herbicide with systemic action for the control of plant species
Triclopyr butoxy ethyl ester and Aminopyralid	A systemic oil-in-water emulsion concentrate herbicide for the control of woody plants as listed for forestry, grass pastures, conservation, and industrial areas

5 Key technological innovations in silviculture

Muedanyi M. Ramantswana and Saulo P. S. Guerra

5.1 Introduction

This chapter will explore key technological innovations that are, or have a high probability of revolutionising silviculture, in South Africa and Brazil. The historical background of technology development will be covered, followed by an examination of the global and local drivers of silviculture technology. The reader will learn about existing and emerging technologies, categorised as machine-specific technical innovations, material input innovations, machine operator-specific innovations, and computerised technology applications, which will be described. While acknowledging that predicting the future is inherently uncertain, this chapter will utilise current knowledge and expert forecasts to simulate possible future scenarios and how technology may evolve in the coming years. This forward-looking analysis will provide valuable insights for decision-makers within the forestry industry, enabling them to proactively plan and strategically adapt to the changes that lie ahead.

5.2 Broad technological advancements

Technology is advancing quickly, and in forestry, innovations are increasingly being integrated into operations such as silviculture re-establishment (regeneration), harvesting and processing (Ramantswana et al., 2021b). Technological advancements must be integrated into forestry operations to improve sustainability through silviculture operations that promote economic viability, environmental friendliness, and social responsibility. New technologies such as big data, the Internet of Things (IoT), artificial intelligence (AI), and sensor technology are becoming more prominent and changing conventional operations across different industries, including forestry.

Even though forestry has lagged in adopting technologies which improve efficiencies, there is now progress in "precision forestry" and the need to improve efficiency and remain sustainable (Choudhry and O'Kelly, 2018). Precision forestry is focused on information and supports economic, environmental, and sustainable decisions by using advanced machines, sensing and analytical tools (Seenivasan and Parthiban, 2017). However, to understand

DOI: 10.4324/9781003391227-5

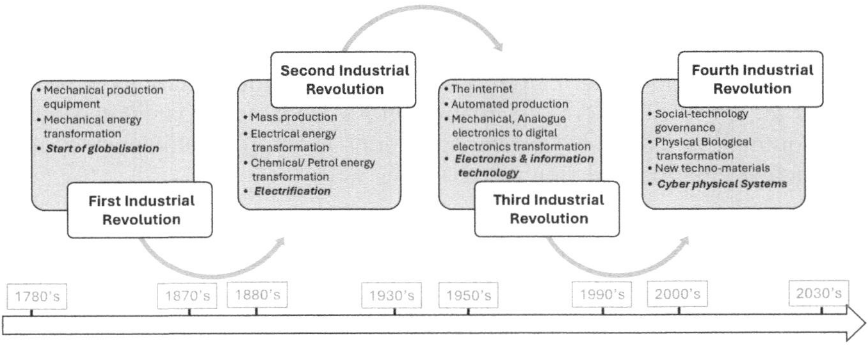

Figure 5.1 The four industrial revolutions – image adapted from (Skilton and Hovsepian, 2018)

current technological developments in forestry and, more specifically, silviculture, one needs to understand the history of technology. The different industrial revolutions experienced by society over the past decades have had a direct impact on technological developments over time; hence, it is essential to understand key inventions that serve as a foundation for our current state (Figure 5.1).

5.2.1 The first industrial revolution

The first industrial revolution is estimated to have commenced around the 18th century (Xu et al., 2018), and it ignited using the steam engine (invented in the 17th century), primarily powered by coal and mechanisation in England (Pozdnyakova et al., 2019, Mohajan, 2019). During this period, there was a significant shift from manual production methods to machines and the increased application of steam and water power (Groumpos, 2021). The steam engine invention ushered in a period of various industries and development, such as textiles, mines, steam-powered railroads, steel production and various other economic activities (Mohajan, 2019). Overall, this first industrial phase increased economic growth, as well as social and political changes. For example, there were improvements in agriculture, transportation, significant goods production, wider global trade, an increase in employment and the introduction of various governing systems (Mohajan, 2019).

5.2.2 The second industrial revolution

The second industrial revolution is a period dated between 1870 and 1914 (Mokyr and Strotz, 1998). The strong emergence of science-based innovation significantly changed productivity and production across various parts of the world (Hull, 1994). This phase is associated with electrification, the invention of internal combustion engines, chemical industries, and increased development of steel (Mohajan, 2019b, Mokyr and Strotz, 1998, Pozdnyakova et al., 2019). During this time, there was a significant increase in industrialised production due to the availability of oil and electricity resources (Xu et al., 2018). Furthermore, many factories expanded, facilitated by the development of assembly line configurations to improve efficiency in mass production (Sharma and Singh, 2020).

5.2.3 The third industrial revolution

The third industrial revolution is dated to have commenced around the mid-1900s after the end of World War II. This revolution is associated with the transition from mechanical and analogue electronic technology to digital electronics (Mohajan, 2021). The drivers of this revolution include improved connectivity, telecommunications infrastructure, large-scale computing and the introduction of the World Wide Web (Skilton and Hovsepian, 2018). This era introduced partial automation, computers and the internet revolution (Groumpos, 2021). Furthermore, this phase also introduced the early shifts to using renewable energy alternatives (e.g. solar, wind, and energy storage technologies, digital manufacturing such as 3D printing) in various manufacturing processes (Troxler, 2013, Heinonen et al., 2015).

5.2.4 The fourth industrial revolution

The fourth industrial revolution builds on the foundation of the third industrial revolution. It is described as the integration of various technologies across the physical, digital, and biological domains. The key technological inventions of this industrial phase involve the use of data, improved computational power, connectivity, analytics, AI, machine learning (ML), robotics, advanced automation, IoT, and biotechnology (Xu et al., 2018, Koc and Teker, 2019).

The significant developments over the past few years have created an environment enabling the forestry industry to adopt new technologies, shifting from manual to machine-based production. The fifth industrial revolution is expected to be dominated by the increased use of collaborative robots to work alongside and assist people when performing work (Ziatdinov et al., 2024). A key theme in the fifth industrial revolution is the interaction between people and technology by leveraging technologies from the fourth industrial revolution, thereby enabling industries like forestry to meet their sustainability goals (Noble et al., 2022).

5.3 Drivers of technology in silviculture

Technology adoption in any field is driven by various factors that influence the level and rate of adoption. As forestry organisations compete in the local and global markets, they must remain competitive and reduce production costs while improving returns to be viable and sustainable in the long term.

Understanding the dynamics of these drivers is important to improving technology adoption decision-making.

5.3.1 Global drivers of technology in silviculture

Silvicultural practices applied within a given land base are dynamic and respond to numerous social, economic, and biological drivers. Silversides (1984) found it concerning at the time that little insight had been focused on silviculture mechanisation, even though he deemed the tasks simpler. Manual labour is a key to a successful silviculture programme (Ketcheson and Smyth, 1977). However, where labour is expensive, it is better to plant trees with machines than to do work by hand (Smith, 1986). Several authors across the globe in countries such as Brazil, Finland, Canada, America, Ireland, New Zealand and South Africa agree that labour shortages and increasing labour costs are driving an increase in mechanisation of silviculture operations (Bayne and Parker, 2012, Betts, 2008, Hallongren et al., 2014, McDermid, 1972, McEwan and Steenkamp, 2014, Van Goor, 1978, Guerra et al., 2019).

Forestry is a complex and dangerous (especially motor manual harvesting) profession compared to alternative manual careers such as roading, construction, and farming. These alternative sectors are considered more attractive because of higher wages, better working conditions, and more opportunities to work on newer, more technologically advanced equipment. Other reasons for labour scarcity include the migration of labour to cities, resulting in a dwindling labour supply for the forestry sectors, which are usually based in rural areas (Bayne and Parker, 2012).

Forestry work involves natural and material risks to the health and safety of forest workers. Forest safety can be influenced by various factors such as forest characteristics, climate, workers' skills, worker methods and technology applied to conduct various activities (McCormack, 2002). Labour conducting silviculture activities generally needs little training, with the main requirements being a "strong back and willingness to work" (Ketcheson and Smyth, 1977); however, with the introduction of modern technologies, the requirements are more stringent. There is a global focus on silviculture practices and the danger they pose to workers' health and safety (Poschen, 1993). Even though motor manual harvesting and other related activities carry the highest safety risk in forestry, re-establishment and tending activities are also considered high risks that may cause health and safety problems. Silviculture workers are at risk of various occupational health issues, some of which are the same or similar to those in harvesting operations, and some are more specific to the task they perform (Trites, 1992). Musculoskeletal problems are associated with carrying out specific, manually oriented work. For example, the manual planting of trees requires prolonged and repetitive non-neural postures, which can cause musculoskeletal problems. Manual planting can be classified as hard continuous work to very hard work (Sullman and Byers,

2000). Furthermore, silviculture crews are exposed to a range of chemicals from pesticides, lubricants, diesel and petroleum fuels, and their emissions (Robinson et al., 1993).

Most silviculture activities are labour-intensive, and through mechanisation, the productivity of these activities can be improved (Silversides, 1984). Furthermore, the need to compete globally may influence the drive to increase productivity through mechanisation to reduce overall costs (Bayne and Parker, 2012). According to Pöry (2014), about two-thirds of silvicultural costs in plantations are incurred during the establishment phase of a plantation. For example, in Brazil, planting between 1,000 and 3,000 seedlings per hectare is typically carried out by a semi-mechanised operation, which is often labour-intensive and time-consuming. This scenario presents an opportunity to increase productivity through technological advancements like planting machines. Conversely, most agricultural crop establishment and management make use of machinery and unmanned robots to improve productivity (Pöry, 2014).

5.3.2 Drivers of silviculture technology development in South Africa and Brazil

In South Africa and Brazil, the main change drivers, which are like those in other countries globally, are summarised in Table 5.1.

Technology and mechanisation of silvicultural tasks cannot be justified in the same way as harvesting. The cost would be the primary driver for silviculture mechanisation, even though all the other reasons, such as quality improvement, productivity increase, reduction in labour requirements, and improving work conditions, were also important (Hodgson, 1979). The employment of forestry labour is governed by the forestry sectoral determination, which sets minimum wages, working hours, number of leave days, and termination rules. In South Africa, the minimum wage has increased significantly over the last decade. The highest increase was in 2013, when the minimum wage in forestry increased by 56% from R7.32 per hour to R11.43 per hour (Labour, 2013). As a result of the increased labour costs, the mechanisation process across harvesting and silviculture was accelerated, even though the industry had been gradually mechanising (Chapman, 2015).

In South Africa, from 1988, most silviculture work was carried out in-house, with 25% being outsourced; in 1998, there was an even balance between in-house silviculture work and outsourced silviculture tasks, but by 2007, about 85% of silviculture work had been outsourced (Längin and Ackerman, 2007). Furthermore, over 60% of plantation forest area is under land claims, which signifies a significant shift from larger private forest growers to smaller forest ownership (Längin and Ackerman, 2007). South Africa is currently undergoing a transition in land ownership, and the silviculture technology systems applied may change depending on the landowner's objectives and economies of scale.

Table 5.1 Summary of current drivers of modernisation

Change drivers	*Sub-factors of change drivers*
Improvements in health and safety	Reduction of exposure to moderate and high risk in operations Reduction of the number of incidents and injuries Improvement of ergonomics and reduction of physically demanding work
Increased operational productivity and reduced costs of operation	Improvement of the overall output of operations Improvement of utilisation and efficiency of tools and equipment Reduction of the costs of performing specific activities
Improved stand productivity	Improvement of the consistency of operations Improved survival of plants Improved uniformity of the stand Increased yield
Environmental aspects/issues	Improved conformance to internationally recognised certification standards Reduced impact on soil, water, and biodiversity was applicable Reduced negative public perceptions
Social aspects/issues	Provision of improved quality of work Better insulation from negative labour and political impacts Reduced absenteeism and labour turnover
Other aspects/issues	Climate change: improved adaptation to unpredictable weather conditions which influence planting seasons and burning regimes Global competition: need to improve and remain globally relevant in terms of costs and overall efficiency

Ramantswana et al. (2021a).

Brink (2001) described the following primary driving forces relevant in South Africa (even though they were derived with a focus on harvesting, they are also applicable to silviculture):

- Political – forest practices, codes, and certification
- Economic – increases fuel cost and cost of capital purchases
- Social – social pressure forcing operations to become more socially acceptable and higher labour costs
- Technical – increased performance of machines concerning technology, resulting in higher productivity and less environmental impacts
- Environment – increased awareness of the environmental impact of machines

Furthermore, McEwan and Steenkamp (2014) have classified the drivers of modernisation of silviculture activities into various interrelated categories. Health and safety, social status (esteem) and education, migration patterns in the South African labour market, HIV AIDS, low productivity, high absenteeism and labour turnover, sectoral wage determination, labour legislation, and social grants were identified as some of the main drivers. Overall, in South Africa, the drivers are mainly labour-related.

Forestry and agriculture are exposed to similar conditions and challenges. A potential driver for silviculture modernisation could be technology developed in agriculture that is migrating to silviculture operations. Advanced equipment and technology exist in agriculture for tilling and tending food crops; it could be that the existing technology is being adapted for forestry silvicultural applications, such as chemical spraying and planting.

5.4 General silviculture trends

General trends within silviculture indicate an increase in the productivity of plantation forests through the application of silvicultural treatments (Evans, 1999). The trending silvicultural features include the application of genetically improved plants, manipulation of stocking outputs to increase total fibre, fertilisation of plants to compensate for nutrient loss, improved site preparation, and establishment practices of organic matter preservation and holistic management of all silvicultural features.

Climate change is a significant factor that will influence forests and associated silviculture practices (Achim et al., 2022, Rubilar et al., 2018). Even though the risks of climate change are known, forest response predictions are still lacking (Szmyt and Dering, 2024). Increasing temperatures, changing rainfall patterns and a possible greater frequency of severe storms, droughts, fires, and pests can all be expected to influence future forests and overall silvicultural practices. What is significant for the forester is how tree distribution, tree survival, tree growth and tree physiology and phenology will be affected (Szmyt and Dering, 2024). The current predictions on forest productivity are based on historic prevailing climatic conditions (Skinner, 2007). However, current management must acknowledge the ongoing changes in the climate to simulate and incorporate the required monitoring and apply relevant adaptation strategies (Gustafson et al., 2020).

The use of precision technology to observe and assess resource availability and constraints to stand productivity and detailed tree attributes of trees in near real time is increasing rapidly (Achim et al., 2022, Rubilar et al., 2018, Qi and Liang, 2024). Responses to various stand treatments (thinning, pruning, fertilising, weeding) or stand damages (wind, drought or storms) can be better understood and quantified using technologies such as remote sensing (Fassnacht et al., 2024). However, the data generated from precision technologies requires considerable preparation and processing for it to be used for modelling and decision-making purposes (Pérez-Carrasco et al., 2022).

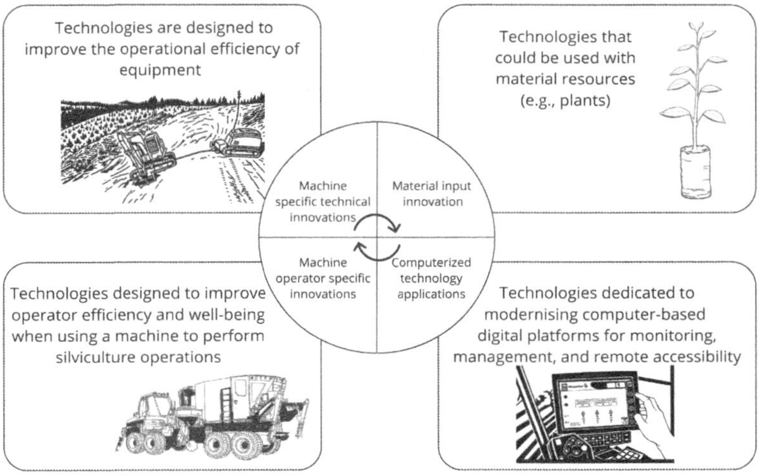

Figure 5.2 Classification of silviculture emerging and future technologies

5.5 Emerging future silviculture technologies

To structure the technologies understandably, the emerging and future technologies in silviculture have been categorised into four broad categories (Figure 5.2), as described by Ramantswana et al. (2021b):

5.5.1 Machine-specific technical innovations

5.5.1.1 Multifunctional machines

Multifunctional machines refer to the integration of traditionally separate re-establishment activities into one or two-pass operations on a machine. Conventional machines in silviculture were designed for a single or dual purpose. For example, a pitting machine only performs pitting, but a tractor with a subsoiler can perform subsoiling and may be integrated with the fertilisation function. A growth in single machines with increased functions can be expected in re-establishment activities. For example, single machines that can prepare the soil, plant, fertilise, and irrigate in a single pass will most likely be adopted in the future. Integrated machines can be more efficient. However, they tend to be more complex and require large carriers to cater to the resources (water, fertiliser, chemical) required for a single-pass planting operation. In the future, smaller carriers and machines may be introduced for small-scale operations, but they will be limited in terms of resources (continuous replenishment). Single- and dual-purpose machines will still be needed in silviculture, as they are more versatile and robust and tend to be more productive when performing single activities.

5.5.1.2 Machine terrain handling enhancements

Machine terrain handling enhancements refer to technologies to enable silviculture equipment to handle poor soil conditions, rougher ground conditions,

Figure 5.3 Winch assist system attached to an excavator performing soil preparation on steep terrain

and steeper slopes. Mechanised harvesting has made significant progress in the development and adoption of technologies for rugged terrain. Technologies that improve machine stability and operator ergonomics are essential to improve safety and efficiencies on steep terrain. In plantation forestry, harvested areas need to be planted as quickly as possible to reduce the temporarily unplanted areas. Existing complementary technologies, which are predominantly used for harvesting equipment, can be adopted in silviculture equipment where the conditions allow. Steep terrain technology, such as self-levelling cabs, bogie quad tracks, and cable assist (Visser and Stampfer, 2015, Visser and Berkett, 2015), can be adopted for silviculture machines performing tasks on steep terrain (Figure 5.3) (Ramantswana, 2023c, Ramantswana, 2023a). Furthermore, silvicultural tools used on ground-based equipment can also be adapted to function optimally on steep terrain. One such solution is the Bracke P12.b planting head, which has a tilt levelling function which enables the planting head to plant upright and well-positioned seedlings on steep slopes, even though the carrier machine may be working on an incline. This ensures good-quality planting of seedlings (Ramantswana, 2023b).

5.5.1.3 Machine automation and robotics

Automation and robotics refer to the machine's ability to perform specific silviculture tasks with minimal human assistance. Automation is described as a technique or operating system to control a process using electronic devices and reducing human intervention to a minimum (Visser and Obi, 2021). Automation can be divided into five categories, namely, operator assistance, partial automation, conditional automation, high automation and driverless

(Lindroos et al., 2017). In forestry, achieving full automation in forest operations will take time, as the working environment is more complex and most areas are remote with limited connectivity. However, there are opportunities for partial automation (e.g. seedling storage and selection for planting (Sobocki et al., 2023)) and the possibility of full automation in the long term. The use of drones in monitoring and spraying activities can be performed (auto navigation) semi-autonomously (Richardson, 2024, Borikar et al., 2022), and this technology is set to expand into the future. In Sweden, there are investigations into the concept of autonomous forest regeneration. The research concept is focused on developing subsystems that will be responsible for enabling autonomous regeneration equipment to be developed in the future (Hansson et al., 2024). Even though the concept was tested, it is not yet well understood, and much work is still needed before a prototype machine can be offered. Some of the challenges to autonomous regeneration are the relation between machine cost and operating speed, sensor robustness in response to vibrations and weather, and precision in detecting the size and type of obstacles during autonomous driving and planting (Hansson et al., 2024).

Robotics focuses on the theoretical and practical application of automated systems. Using big data and advanced sensors, robots will be able to evaluate conditions and make decisions based on prevailing conditions. Furthermore, there are activities performed by machines, for example, tray filling, seeding, and irrigation, based on sensor technology and configuration. There are more opportunities for advanced automation in the nursery process, where planting materials like cuttings can be automatically cut, selected and planted with minimal human intervention (Hendricks, 2024).

5.5.1.4 Application of unmanned aerial vehicles (UAVs) to monitor and conduct silviculture activities

UAVs or drone technology is useful for field mapping, stand assessment, monitoring and executing silviculture activities such as spraying with precision (Figure 5.4). This technology has rapidly been adopted in South Africa and is expected to continue expanding rapidly in the future as the technology is improved and accompanying sensors are improved. UAVs have revolutionised practices in the forestry industry because forest stands can be assessed quickly and regularly for issues such as water stress, pests, diseases, and stand damage (drought, hail, frost, wind, fire). Furthermore, the use of digital aerial photogrammetry combined with objective detection models has presented opportunities to improve data collection and analysis in forest stands. In South Africa, this technology has been used for detecting young pine trees, stand uniformity and weed detection (Kok et al., 2025, Jordaan et al., 2025)

UAVs are versatile and productive tools which can be used to execute certain silviculture tasks, such as pesticide spraying before and after planting. When compared to conventional methods such as manual and semi-mechanised

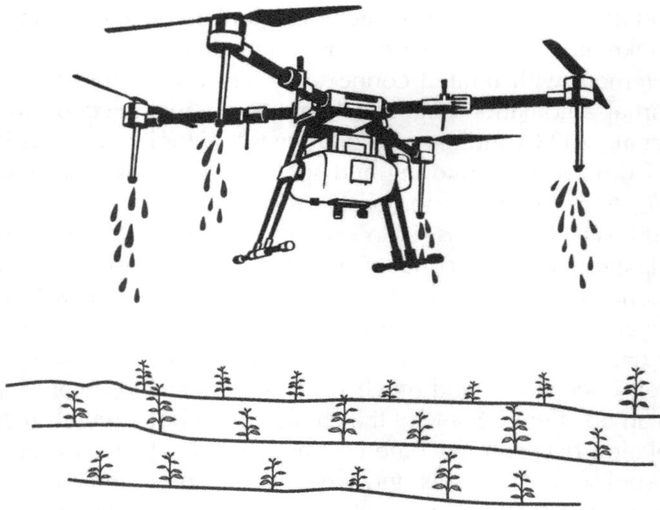

Figure 5.4 AUV spraying pesticide on stand after planting

methods, UAVs are significantly more productive. For details about the application of UAVs in performing silviculture operations, refer to Sections 3.4.3 and 4.1.4.3. Regarding future developments, Vergouw et al. (2016) predict that UAVs will become more efficient in terms of size and improved battery life. Furthermore, they will become more autonomous, wherein they will be able to determine flight routes, sense and avoid collisions, and adapt to changing weather conditions. Finally, UAVs will be able to operate in swarms, wherein a few drones may be able to work together and communicate to complete a task (Vergouw et al., 2016).

5.5.1.5 Machine self-diagnosis and maintenance

Self-diagnosis encompasses reconfiguration, troubleshooting, and self-maintenance with very little human input. The forestry environment is harsh, and forestry equipment is subject to wear, tear, and repairs. Machine availability is an essential aspect of efficiency in forestry activities. Therefore, the prevention of breakdowns or the quick repairs of machines is critical for viability. Prediction of maintenance requirements can be done through the collection and analysis of data received from the machine sensors. Big data analytics offer an opportunity to predict machine failures before they happen or to thoroughly diagnose the issue with the machine and provide adequate information to facilitate the speedy repair of a machine that has a breakdown (Maktoubian et al., 2021). Remote pre-diagnosis can gather data about the machine's fault before a technician visits the site to conduct the repair (Spencer and Torres, 2022). This can save costs and ensure the technician is adequately equipped to resolve the issues. In future, digitised communication protocols integrated into machines will facilitate predictive maintenance and reduce downtime by facilitating accurate and problem-based repairs, thereby reducing the need to diagnose the issue physically on site.

5.5.2 Material inputs innovation

5.5.2.1 *Paper-based plant pots*

In South Africa, containerised trays with degradable paper pots have increased compared to conventional filled insert tray systems (refer to Section 2.3.2.1 for details of paper pots in the nursery process). Paper pots are paper-wrapped plugs which make use of air pruning to change the root formation of seedlings/cuttings and improve the desired establishment traits (Gilgunn, 2024, Biggs, 2024).

In Brazil, the pursuit of developing biodegradable tubes has intensified, as they can reduce operational processes and costs, besides not generating waste for the environment and lowering the carbon footprint (da Silva Santana et al., 2019), since they can degrade more quickly in the soil (Guerra et al., 2017).

When compared to plastic insert trays, paper pots are described as having a more consolidated root plug (5.5), better moisture retention, more active white roots, more robust and suited for semi-mechanised and mechanised planting and more efficient to plant (less handling) (Gómez et al., 2019).

In Brazil, the use of water collection plastic trays has become a practical option to replace activities related to arranging seedlings in the tray, known as alternating, as they have designated spots for inserting seedlings. This approach enhances operational efficiency and allows for better water and fertiliser utilisation (Gómez et al., 2019). According to Cury et al. (2024), the use of biodegradable tubes, water collection trays, and seedling alternating methods in trays are options to achieve increased production efficiency, combined with cost reduction.

Early results in South Africa indicate that paper pot plugs performed better in semi-mechanised planting operations when considering survival, growth and uniformity when compared to the conventional Unigro 128 plugs (Hechter, 2024). Further research and field tests of paper pots across various sites are necessary to quantify their effects accurately. In future, the adoption of paper pot systems may increase based on their suitability to modern planting systems (Figure 5.5).

5.5.2.2 *Nanofertilisers and fertiliser tablets*

Fertilisation is applied to correct nutritional deficiencies on the site and improve tree growth. In South Africa, comprehensive research focused on fertilisation strategies to optimise yield (i.e. fertilisation accompanied by

Figure 5.5 Consolidated root plug with paper pot

weeding and application of fertilisers on responsive sites) (Brink, 2012). However, very little work has been published on efficient technologies for fertiliser application. A similar scenario occurs in Brazil, where the nano-fertiliser is an emerging technology, with promising results in crops (e.g. corn and beans), where chitosan nanoparticles containing nitrogen, phosphorus and sodium enhanced the crop productivity (Lima and Roggia, 2022)

Current fertilisation methods of applying granular fertiliser are manually intensive (placements in slots or holes 25–30 cm away, 10 cm depth) and require a lot of handling (measuring small quantities) (Hans, 2013). In Brazil, the most common method of applying granular fertiliser is semi-mechanised, combined with subsoiling. Placements in the subsoiled line consist of 70% of the dose at 15–25 cm depth and 30% of the dose at 40–60 cm depth.

The use of slow-release fertiliser tablets, which are placed below the seedling root plug at planting, can improve efficiencies in fertiliser application operations. The use of fertiliser tablets in areas where spot cultivation is practised extensively will become increasingly important in the future. However, integrating tablet technology into continuous soil cultivation would be complex. The use of fertiliser tablets has shown an increase in growth in some species, although not at the level of nitrogen–phosphorus applications. This technology will be an important alternative for operations where fertiliser has to be applied at planting (Neilsen, 1996).

Nanotechnology in forestry is an emerging field with the potential to grow in the future. In silviculture, nano fertilisers have the potential for future adoption in nutrient management applications (Singh et al., 2021). Nanoparticles are nanoscale fertilisers prepared in the form of fine particles with nanoscale dimensions. By changing the composition of fertiliser based on nutrient requirements for specific species, nano fertilisers can be applied to forestry plants (Marimuthu et al., 2017). The advantage of using these fertilisers is that they can easily be controlled and targeted to specific crops to enhance tree growth with minimal waste (Kanjana, 2017, Kumar et al., 2025). Although this technology has not yet been adopted in the South African and Brazilian forestry industry, it has potential once it has been widely tested and the effects have been quantified.

5.5.2.3 Ultra-low emission engines

Globally, there is a need to reduce engine emissions due to climate change. Forestry vehicles need to adhere to emissions standards for each country, and carbon emissions can be reduced by using environmentally friendly engine technologies (Cosola et al., 2016). Existing engines are expected to be refined (Tiers 1–5, Stages I–V) to lower or eliminate air pollutant emissions as the globe aims to meet sustainability goals. Up to now, the two main strategies that have been applied to reduce engine emissions include in-cylinder approaches and exhaust after-treatment device engines (Dallmann and Menon, 2016). As emissions regulations are enforced, vehicles (tractors, dozers, excavators, forwarders) used to perform silviculture operations will need to comply with the standards. There are various alternatives to diesel engines, such as biofuels, hydrogen for fuel cells, electrification, and hybridisation (Mergl et al., 2021, Ettl et al., 2014, Kreis and Fischer, 2024, Ramantswana, 2022).

5.5.3 Machine operator-specific innovations

5.5.3.1 Ergonomically friendly cabs

The machine cab is the working environment for the operator. Existing cabs still possess significant gaps in terms of ergonomic design. Cabs can pose health and safety risks (e.g. musculoskeletal disorders) to the operator if poorly designed or ergonomic practices are ignored. In South Africa and Brazil, ergonomic aspects related to silviculture machines have not been studied comprehensively. Non-purpose-built equipment used in forestry is adopted from agriculture (e.g. tractors) or construction (e.g. excavators), and it is not tested for ergonomic suitability in forestry conditions.

However, in Brazil, the cabins of agricultural tractors and construction excavators comply with Brazilian Standards for Agricultural and Construction Machinery Cabins, such as: NR 12 – Safety at Work in Machinery and Equipment; NR 17 – Ergonomics; NR 31 – Safety and Health at Work in Agriculture, Livestock, Forestry, and Aquaculture; ABNT NBR ISO 26322-1 – Tractors for Agriculture and Forestry – Safety – Part 1: Standard Tractors; ABNT

NBR ISO 4252 – Agricultural Tractors – Operator's Workplace, Access, and Exit – Dimensions; ABNT NBR ISO 4253 – Agricultural Tractors – Operator Seat Accommodation; ABNT NBR ISO 4254-1 – Agricultural Machinery – Safety – General Requirements; ABNT NBR ISO 5700 – Roll-Over Protective Structures (ROPS) – Static Test Method and Acceptance Conditions; ABNT NBR ISO 3449 – Falling-Object Protective Structures (FOPS); ABNT NBR ISO 6682 – Earth-Moving Machinery – Zones of Comfort and Reach for Controls; ABNT NBR ISO 2631 – Evaluation of Human Exposure to Whole-Body Vibration; ABNT NBR ISO 5721-1 – Agricultural Tractors – Operator's Field of Vision.

Silviculture machines operate (e.g. soil preparation machines) in harsh environments (rough terrain), and operators tend to adopt poor work postures, and they are exposed to high temperatures in the cab when it is hot (Gellerstedt, 2000). Ergonomically friendly work environments comprise cabs with improved lighting, low vibration, temperature control options, good access to the cab, adequate cab dimensions, good visibility, and enhanced range of motion and positioning (Machado et al., 2020). As silviculture machines are developed and adopted, it can be expected that existing and new technologies such as rotating and levelling cabs, adjustable seats (accommodating variations in individuals), seat heaters (warming of cold seat to required temperature), adjustable armrests, adjustable cab suspension, turnaround seats with controls and LED lighting will be adopted (Gellerstedt, 1998, Ponsse, 2024, Tigercat, 2022, Komatsu, 2025). These advancements in cabs, combined with partial automation of some functions, will improve operator comfort, improve working posture, and prevent stress on the operator's body.

5.5.3.2 *Simulation training*

Forestry simulators offer an opportunity for new operators to learn the basics of how to operate a machine cost-effectively, and for existing operators to be assessed and upskilled (Schwegman et al., 2023, Burk et al., 2023). Currently, silviculture machine operators learn how to operate machines through hands-on training using the actual machine. This is costly in terms of loss in production and potential damage to the machine through learning. The lack of silviculture machine-specific training has been driven by the fact that most silviculture equipment has been adopted from other industries (e.g. forwarder machine from harvesting, tractor from agriculture), and operators used are usually already familiar with operating the base machines. As silviculture machines become more purpose-built, more costly, and technologically advanced, simulator training will become more crucial for operators to use the machines effectively. Simulator training will be beneficial because it is more cost-effective, provides a safe working environment, makes it easier to monitor progress and makes skills transfer occur between simulated and real forest environments (Burk et al., 2023). Equipment manufacturers, grower companies, and researchers will need to collaborate and simulate silviculture environments and activities to improve machine operator performance.

5.5.4 Computerised technology applications

5.5.4.1 Artificial intelligence

According to Shivaprakash et al. (2022), AI is a wide-ranging branch of computer technologies concerned with building intelligent machines capable of augmenting, automating, and accelerating key day-to-day tasks associated with human intelligence. It involves extracting patterns, predicting "future states", and detecting anomalies. AI in forestry applications is enabled by technologies such as sensors, drones, and satellites, which facilitate data collection to improve understanding and provide accurate information for decision-making (Figure 5.6). AI systems can be trained to simulate human thinking and optimise decision-making. Data gathered through advanced imagery can be processed, prepared for analysis, feature extraction, model training and prediction application systems (Holzinger et al., 2024).

AI involves a broader focus on machines performing tasks that would otherwise require human intelligence. ML, a subset of AI, uses algorithms that self-learn from input data. Deep learning (DL), a further subset, makes use of multi-layered neural networks to enhance the accuracy of predictions

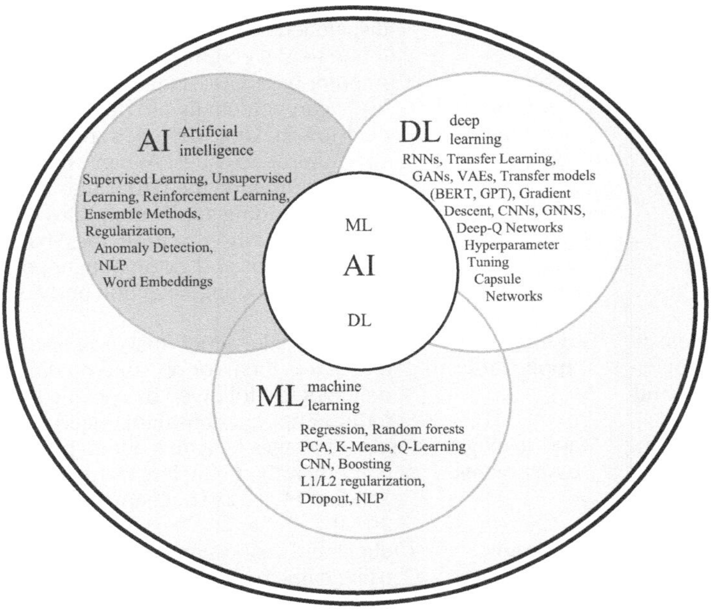

Figure 5.6 Illustration of the relation and differences among artificial intelligence (AI), machine learning (ML), and deep learning (DL) (Wang et al., 2025)

and decision-making (Wang et al., 2025). AI can be used in forestry applications such as tree inventory management, silviculture, harvesting, and road management. Some existing applications of AI in silviculture operations are described in Table 5.2.

As more data is collected and made available from silviculture operations, the use of AI systems is expected to increase exponentially.

5.5.4.2 Real-time machine and operator performance monitoring

Acquiring data from machines is crucial for monitoring performance. Silviculture machines can be equipped with hardware and software that provide comprehensive information about the machine (Global Navigation Satellite System – GNSS, such as GPS technology) (Guimier and Heidersdorf, 1998). Telematic systems can track the performance of machines operating at remote locations, provide information about the progress of machines working on site, and estimate completion times. Furthermore, information such

Table 5.2 Applications of AI in silviculture operations

Technology	*Application*	*Descriptions*
Computer vision	Plant quality assessments in nurseries	Cameras are used to take images, which are used to assess plant quality attributes. Assists in the selection of good-quality plants (based on plant quality index) before being dispatched for planting to enhance uniformity in the field (Peerbhay and Maresch, 2025)
Telematic systems	Machine operator assistance	Computer-based assistance systems underpinned by AI can potentially aid operators in decision-making when performing various silvicultural activities. AI can assist operators with determining the best routes to optimise travel, providing application rates of materials (fertiliser or pesticides) based on site conditions and automating repetitive tasks (e.g. seedling detection and automatic irrigation)
Remote sensing technologies (satellite and AUVs)	Fertiliser application	Drones can take aerial images of tree crowns and assess them for possible nutrient deficiencies, followed by specific remedial fertiliser applications (Holzinger)
Remote sensing technologies (satellite and AUVs)	Stand damage assessments	Assess damages to stands due to pests, diseases, and other abiotic factors (Nzuza et al., 2025, Sandino et al., 2018, Giannakidou et al., 2024)
Weeding technologies	Advanced weeding	Optical and vision systems for weed sensing by using cameras, image processing algorithms and analysis to optimise the application of resources such as chemicals to target specific weeds (Pajares et al., 2013, Wu et al., 2021)

as material usage (e.g. fuel, oil), service and maintenance requirements, and engine and operator performance level reports can also be generated (Henny, 2022). This data can be transmitted via satellite or cellular service (where connectivity allows). The available data can be accessed in real time via a computer dashboard or smartphone (CAT, 2025).

5.5.4.3 Processing and application of big data

Machines (e.g. telematic systems) and monitoring systems (e.g. remote sensing data) generate large quantities of data daily. Data generation is expected to continue to increase in the future as value is derived from existing data (Zou et al., 2019). Broadly, big data is a component of precision forestry, which also includes the use of cloud computing, IoT, and mobile internet. Some examples of big data that can be generated from silviculture and related operations include productivity information, pitting and planting positions, and aerial imagery captured frequently to monitor stand health and assess damage on forest stands. Once processed, big data provides opportunities to improve efficiencies in silviculture activities.

5.5.4.4 Remote control of operating machines

Remote control of silviculture machines is a potential future possibility. Radio-controlled devices can remotely operate machines that perform a specific task without the operator being present in the cab. This removes the need for an operator to be inside the cab for the machine to execute its function. The adoption of remote-controlled silviculture machines will be driven by the need to improve health and safety (i.e. reduce whole-body vibrations experienced by the operator in the cab, the operator is protected from dust, noise or safety risks prevalent at the work site), the need for lighter machines (e.g. heavy machines cause compaction and rutting) on the site, the need to improve productivity (reduced daily commuting to remote sites) and enhance recruitment of machine operators (Forest Machine Magazine, 2024, SCA, 2024). Investigations into the use of remote-controlled silviculture machines are in the infancy stage. A collaborative research group is testing a remotely controlled soil scarification machine in Sweden; this groundwork will serve as a foundation for future work in this area (SCA, 2024). Although there are prospects for the adoption of remote-controlled silviculture machines in the future, the lack of infrastructure (stable communication signals) and high initial costs may be limiting barriers in the short term.

5.6 Overview of the cost of modernisation

As the forestry industry adopts various instruments to improve operational efficiencies combined with better yields, the actual costs need to be defined and quantified (Figure 5.7). In South Africa and Brazil, grower companies

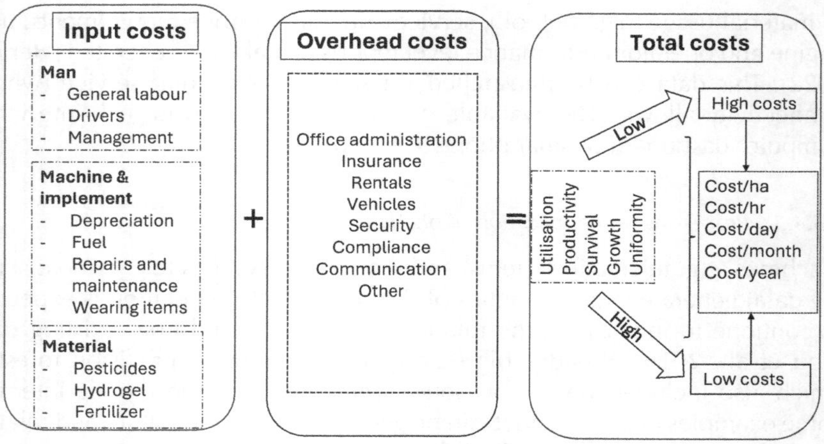

Figure 5.7 Flow diagram of operational silviculture cost outputs

can utilise their own resources (in-source), outsource, or employ a combination of both approaches. The most common approach in both countries is the use of contractors to perform silviculture work. When silviculture work is outsourced, the grower company and contractor agree on rates (price) for performing the work based on the productivity, complexity of the work, magnitude of work, and costs of resources required to perform the work.

Contractors typically charge a rate to perform the work. The rate can be per hectare, per man-day, per activity, per hour, or per day. This arrangement can vary depending on the circumstances. Contractors are usually appointed for periods ranging from one year to five years, even though there can be exceptions. The contractor must determine the correct rate to ensure the business is viable in the long term. This cost determination needs to be based on actual costs or accurate assumptions. The details of cost determination in silviculture operations are beyond the scope of this book; however, the main cost categories in silviculture operations comprise fixed, variable, and overhead costs. In manually oriented operations, the labour costs are the highest cost component; in mechanised operations, the machine costs (capital and fuel costs) are the highest cost components. In mechanised operations, low machine utilisation and productivity result in increased costs. Furthermore, when tree survival, growth, and uniformity are poor, it also leads to high costs; the inverse leads to lower costs. Productivity is a crucial factor in mechanised operations, as it directly impacts costs; therefore, operational productivity norms must be established and met to ensure business viability.

a The cost of silviculture equipment and implements used seasonally is complex, as it depends on variable assumptions. Industry needs robust costing models that forestry stakeholders can use to accurately quantify the cost per ha, cost per hour, and cost per month/year of utilising various technologies. Given that most silviculture work is outsourced, grower companies and contractors must collaborate in recording, reviewing, and refining operational costs.

b The whole is greater than the sum of its parts. Optimising one operation without considering its impact across the entire value chain is counterproductive. Foresters must assess the "cost" of modernising a single operation on downstream activities, focusing on safety, productivity, environment, quality, and social implications. For example, if mulching is adopted, it is essential to understand its effects on soil properties (biological, physical, and chemical properties), tree survival and growth, access for other site activities, and fire risks and quantify it in financial terms.

c What are the actual gains and losses? Not every modernised system will bring significant improvements; some studies have shown that using mechanised systems may lead to poor survival compared to conventional manual systems. New technologies must be thoroughly investigated, and knowledge gaps must be filled quickly for forestry stakeholders to make informed decisions.

d We need to better understand silviculture mechanisation's impact on site productive capacity, with a particular focus on organic matter, nutrient availability, soil disturbance, soil compaction, and soil loss. What are the short-term and long-term implications of our current operations? Accessible guidelines and decision-support tools are needed to equip managers with the relevant information.

5.7 Summary

This chapter explored key technological innovations in silviculture, focusing on those currently relevant and potentially significant in the future, particularly in South Africa and Brazil. This chapter reviewed the drivers of technology adoption in silviculture, linking them to the industrial revolutions, and discussed general silviculture trends. Moreover, this chapter covered machine-specific innovations, such as multifunctional machines and terrain handling enhancements, alongside automation and robotics, including the use of UAVs. It also examined material input innovations, such as the adoption of paper-based plant pots, and operator-specific innovations, including ergonomically friendly cabs and simulation training. Finally, it also explored computerised technology applications, including AI, big data processing, and remote control of machines.

Review questions

1 What are the global drivers of silviculture technology developments?
2 Name and explain examples of current or forecasted machine-specific technologies that we can expect to be adopted in silviculture operations

3 Name and explain examples of current or forecasted material input innovations that we can expect to be adopted in silviculture operations
4 Name and explain examples of existing or forecasted operator-specific innovations in silviculture operations we can expect to be adopted in silviculture operations
5 Name and explain examples of current or forecasted computerised technology applications that we can expect to be adopted in silviculture operations

References

Achim, A., Moreau, G., Coops, N. C., Axelson, J. N., Barrette, J., Bédard, S., Byrne, K. E., Caspersen, J., Dick, A. R. & D' Orangeville, L. 2022. The changing culture of silviculture. *Forestry,* 95, 143–152.

Bayne, K. M. & Parker, R. J. 2012. The introduction of robotics for New Zealand forestry operations: Forest sector employee perceptions and implications. *Technology in Society,* 34, 138–148.

Betts, J. 2008. Recent workforce trends and their effects on the silviculture program in British Columbia. *Ecosystems,* 1, 12.

Biggs, S. 2024. Growing seedlings for current and future planting technologies. *Modern Silviculture Symposium*. Howick: Nelson Mandela University.

Borikar, G. P., Gharat, C. & Deshmukh, S. R. 2022. Application of drone systems for spraying pesticides in advanced agriculture: A review. *IOP Conference Series: Materials Science and Engineering*. IOP Publishing, 012015.

Brink, M. 2001. *Development of a Method to Forecast Future Systems in the Forest Engineering Value Chain*. Stellenbosch: Stellenbosch University.

Brink, M. 2012. Certification as a risk management strategy. *In:* Bredenkamp, B. & Upfold, S. (eds.) *South African Forestry Handbook*. Pretoria: South African Institute of Forestry Research.

Burk, E., Han, H.-S., Smidt, M. & Fox, B. 2023. Effectiveness of simulator training compared to machine training for equipment operators in the logging industry. *International Journal of Forest Engineering,* 34, 373–384.

CAT. 2025. *Telematics – Your Link to Equipment Cost Savings* [Online]. CAT.com: CAT. Available: https://www.cat.com/en_US/articles/ci-articles/forestry-telematics.html [Accessed 2 May 2025].

Chapman, C. 2015. Mechanisation changing the face or forestry. *SA Forestry*. Hillcrest: Green Forest Media.

Choudhry, H. & O' Kelly, G. 2018. Precision forestry: A revolution in the woods. *Advanced Technologies Could Improve Forest Management Significantly. What Areas Are Most Promising, and How Can Forestry Companies Start Their Digital Transformation?* Singapore: McKinsey&Company.

Cosola, G., Grigolato, S., Ackerman, P., Monterotti, S. & Cavalli, R. 2016. Carbon footprint of forest operations under different management regimes. *Croatian Journal of Forest Engineering: Journal for Theory and Application of Forestry Engineering,* 37, 201–217.

Cury, L. D., da Silva, L. F., de Oliveira, V. C., de Mello Arthur, M., Rodrigues, S. A., da Silva, J. M. S. & Guerra, S. P. S. 2024. Crescimento de mudas de Eucalyptus urophylla produzidas em bandejas coletoras de água e convencional nos sistemas de tubetes degradáveis e polipropileno. *Ciência Florestal,* 34, e73734.

da Silva Santana, J. A., da Costa, T. L. N., da Silva, B. R. F., Jr, V. C. B. J., do Prado Costa, M. & do Canto, J. L. 2019. Utilização de recipientes biodegradáveis de bambu no desenvolvimento de mudas de Enterolobium contortisiliquum (Vell.) Morong. *Brazilian Journal of Development,* 5, 15912–15921.

Dallmann, T. & Menon, A. 2016. *Technology Pathways for Diesel Engines Used in Non-Road Vehicles and Equipment.* Washington: International Council on Clean Transportation (ICCT).

Ettl, J., Thuneke, K., Remmele, E., Emberger, P. & Widmann, B. 2014. Future biofuels and driving concepts for agricultural tractors. *European Biomass Conference and Exhibition*. Hamburg, Germany, 23–25.

Evans, J. 1999. Sustainability of forest plantations: A review of evidence and future prospects. *The International Forestry Review,* 1 (3), 153–162.

Fassnacht, F. E., White, J. C., Wulder, M. A. & Næsset, E. 2024. Remote sensing in forestry: Current challenges, considerations and directions. *Forestry: An International Journal of Forest Research,* 97, 11–37.

Forest Machine Magazine 2024. Remote-controlled scarification machine underway. *Forest Machine Magazine*. Glasgow: Forest Machine Magazine.

Gellerstedt, S. 1998. A self-leveling and swiveling forestry machine cab. *Journal of Forest Engineering,* 9, 7–16.

Gellerstedt, S. 2000. Ergonomic guidelines for forest machines. *Proceedings of the Human Factors and Ergonomics Society Annual Meeting*. Los Angeles, CA: SAGE Publications, 477–480.

Giannakidou, S., Radoglou-Grammatikis, P., Lagkas, T., Argyriou, V., Goudos, S., Markakis, E. K. & Sarigiannidis, P. 2024. Leveraging the power of internet of things and artificial intelligence in forest fire prevention, detection, and restoration: A comprehensive survey. *Internet of Things,* 26, 101171.

Gilgunn, M. J. 2024. *The Development, Root Egress, and Root Growth Potential of Ponderosa Pine Grown in Paper Pots and Air Pruning Trays.* Moscow, ID: University of Idaho.

Gómez, JV, Maciel, J. & Silva, M. 2019. Bandejas coletoras x bandejas tradicionais na produção de mudas de Eucalyptus grandis Hill ex Maiden x Eucalyptus urophylla ST Blake. Silviculture for Eucalyptus plantations. *Congress Themes: C. Forests and Forest Products for a Greener Future*. Resumos. Curitiba. 7.

Groumpos, P. P. 2021. A critical historical and scientific overview of all industrial revolutions. *IFAC-PapersOnLine,* 54, 464–471.

Guerra, M. S., Barbosa, M. S., Costa, E. & Vieira, G. H. C. 2017. Recipiente biodegradável e substratos para mudas de maracujazeiro. *Revista de Agricultura Neotropical,* 4, 50–54.

Guerra, S. P. S., Soler, R. R., Sereghetti, G. C. & Oguri, G. 2019. An evaluation of the economics and productivity of fully mechanised tree seedling planting in Brazil. *Southern Forests: A Journal of Forest Science,* 81, 281–284.

Guimier, D. Y. & Heidersdorf, E. 1998. Forestry operations into the next century. *The Forestry Chronicle,* 74, 68–71.

Gustafson, E. J., Kern, C. C., Miranda, B. R., Sturtevant, B. R., Bronson, D. R. & Kabrick, J. M. 2020. Climate adaptive silviculture strategies: How do they impact growth, yield, diversity and value in forested landscapes? *Forest Ecology and Management,* 470, 118208.

Hallongren, H., Laine, T., Rantala, J., Saarinen, V.-M., Strandström, M., Hämäläinen, J. & Poikela, A. 2014. Competitiveness of mechanized tree planting in Finland. *Scandinavian Journal of Forest Research,* 29, 144–151.

Hans, R. R. 2013. *Initial Growth Responses to Controlled Release Fertilizer Application at Establishment of Commercial Forestry Species in South Africa.* Stellenbosch: Stellenbosch University.

Hansson, L. J., Sten, G., Rossander, M., Lideskog, H., Manner, J., Van Westendorp, R., LI, S., Eriksson, A., Wallner, A. & Rönnqvist, M. 2024. Autoplant—Autonomous site preparation and tree planting for a sustainable bioeconomy. *Forests,* 15, 263.

Hechter, U. 2024. Testing paper pots and plastic inserts for a Eucalyptus clone and seedling on a poor site at Dumbe. *Modern Silviculture Symposium.* Howick: Nelson Mandela University.

Heinonen, S., Karjalainen, J. & Ruotsalainen, J. 2015. Towards the third industrial revolution. *Findland Future Research Centre FFRC EBook,* 6, 2015.

Hendricks, J. 2024. Automation in forestry nurseries: Can you afford to stay behind. *Modern Silviculture Symposium – Silviculture 4.0: Smart Technology from Nursery to Field (15–17 October 2024, Howick, South Africa).* Howick: Nelson Mandela University.

Henny, Z. 2022. *Telematics Systems in Silviculture Equipment* [Online]. www.forsilvitech.com: Forsilvitech. Available: https://www.forsilvitech.com/article/steep-terrain-solutions-for-mechanised-ground-based-silviculture-equipment/ [Accessed 2 May 2025].

Hodgson, T. 1979. Justifications for mechanisation in silviculture. *South African Forestry Journal,* 110, 16–19.

Holzinger, A., Schweier, J., Gollob, C., Nothdurft, A., Hasenauer, H., Kirisits, T., Häggström, C., Visser, R., Cavalli, R. & Spinelli, R. 2024. From industry 5.0 to forestry 5.0: Bridging the gap with human-centered artificial intelligence. *Current Forestry Reports,* 10, 442–455.

Hull, J. P. 1994. The second industrial revolution and the staples frontier in Canada: Rethinking knowledge and history. *Scientia Canadensis,* 18, 22–37.

Jordaan, M., Peerbhay, K. & Talbot, B. 2025. Assessing stand heterogeneity in plantations using UAV-derived data and deep learning for crown segmentation. *In:* Erasmus, J., Talbot, B., Dlokweni, P. & Grube, G. (eds.) *Precision Forestry: At the Dawn of Artificial Intelligence (AI),* 3–7 February 2025. Stellenbosch: Department of Forest and Wood Science, Faculty of AgriSciences, Stellenbosch University, 72.

Kanjana, D. 2017. Advancement of nanotechnology applications on plant nutrients management and soil improvement. *In:* Prasad, R., Kumar, V., Kumar, M. (eds) *Nanotechnology: Food and Environmental Paradigm.* Singapore: Springer, 209–234.

Ketcheson, D. & Smyth, J. 1977. Labor: The key to Ontario's silvicultural program. *The Forestry Chronicle,* 53, 336–340.

Koc, T. C. & Teker, S. 2019. Industrial revolutions and its effects on quality of life. *PressAcademia Procedia,* 9, 304–311.

Kok, J., Ackerman, S., Talbot, B. & Erasmus, J. 2025. Assessing stand heterogeneity in plantations using UAV-derived data and deep learning for crown segmentation. *In:* Erasmus, J., Talbot, B., Dlokweni, P. & Grube, G. (eds.) *Precision Forestry: At the Dawn of Artificial Intelligence (AI),* 3–7 February 2025. Stellenbosch: Department of Forest and Wood Science, Faculty of AgriSciences, Stellenbosch University, 72.

Komatsu. 2025. Sweden: Komatsu. Available: https://www.komatsuforest.com/explore/ergonomics/from-a-technical-perspective [Accessed 2 May 2025].

Kreis, A. & Fischer, J. 2024. Towards sustainable forest management–Advancing autonomy approach in forestry operations for the forests of the future. *9th International Engineering, Sciences and Technology Conference (IESTEC), 2024*. IEEE, 599–603.
Kumar, S., Sharma, Y., Khandelwal, V., Rawat, K. & Patil, A. 2025. Applications of nanotechnology in fertilizers: A review study. *Sustainable Chemistry for the Environment,* 100247.
Labour, D. O. 2013. Basic conditions of employment Act, No 75 of 1997 Sectorial Determination 12, Forestry Worker Sector, South Africa. *In:* Labour (ed.) Pretoria: Department of Labour.
Längin, D. & Ackerman, P. 2007. *South African Forest Engineering Survey*. Stellenbosch: Stellenbosch University and ICFR.
Lima, M. & Roggia, I. 2022. Inserção da Nanotecnologia como ferramenta na produção Agrícola: relato de atividade facilitadora de acesso a novas tecnologias. *Disciplinarum Scientia| Ciências Humanas,* 23, 15–30.
Lindroos, O., La Hera, P. & Häggström, C. 2017. Drivers of advances in mechanized timber harvesting–A selective review of technological innovation. *Croatian Journal of Forest Engineering: Journal for Theory and Application of Forestry Engineering,* 38, 243–258.
Machado, J. S., Schettino, S., Colen, F. & Pereira, S. 2020. Ergonomic evaluation of agricultural tractors used in forestry operations. *Engenharia na Agricultura,* 29 (1), 1–12. https://doi.org/10.13083/reveng.v29i1.8834.
Maktoubian, J., Taskhiri, M. S. & Turner, P. 2021. Intelligent predictive maintenance (Ipdm) in forestry: A review of challenges and opportunities. *Forests,* 12, 1495.
Marimuthu, S., Gunasekaran, K. & Subramanian, K. 2017. Applications of nanotechnology in forestry. *Forestry Technologies-A Complete Value Chain Approach,* 20.
Mccormack, R. 2002. Safety and occupational health in forestry operations in Australia-changes in approach through time. *In: FAO RAP Publication 2002/14*. Bangkok: Food and Agriculture Organization of the United Nations, 193–198.
Mcdermid, R. W. 1972. Mechanization, safety, and manpower in southern forestry: 21st annual forestry symposium. *Proceedings [of the] Annual Forestry Symposium,* Baton Rouge, LA, 1.
Mcewan, A. & Steenkamp, J. 2014. Silviculture modernization in the South African forestry industry. *Proceedings of the Second International Congress of Silviculture, 26th–29th November,* Florence. Florence: Accademia Italiana di Scienze Forestali.
Mergl, V., Pandur, Z., Klepárník, J., Kopseak, H., Bačić, M. & Šušnjar, M. 2021. Technical solutions of forest machine hybridization. *Energies,* 14, 2793.
Mohajan, H. 2019. The first industrial revolution: Creation of a new global human era. *Journal of Social Sciences and Humanities,* 5 (4), 377–387.
Mohajan, H. 2021. Third industrial revolution brings global development. *Journal of Social Sciences and Humanities*, 7 (4), 239–251.
Mokyr, J. & Strotz, R. H. 1998. The second industrial revolution, 1870–1914. *Storia dell'economia Mondiale,* 21945.
Neilsen, W. 1996. Response of Eucalyptus nitens and Eucalyptus regnans seedlings to application of various fertilisers at planting or soon after planting. *New Zealand Journal of Forestry Science,* 26, 355–69.
Noble, S. M., Mende, M., Grewal, D. & Parasuraman, A. 2022. The Fifth Industrial Revolution: How harmonious human–machine collaboration is triggering a retail and service [r] evolution. *Journal of Retailing,* 98, 199–208.

Nzuza, P., Michelle Schröder, Heim, R., Daniels, L., Slippers, B., Hurley, B., Germishuizen, I., Sivparsad, B., Roux, J. & Maes, W. 2025. Assessing Gonipterus sp. n. 2 Defoliation Levels Using Multispectral Unmanned Aerial Vehicle (UAV) Data in Eucalyptus Plantations. *In:* Erasmus, J., Talbot, B., Dlokweni, P. & Grube, G. (eds.) *Precision Forestry: At the Dawn of Artificial Intelligence (AI)*, 3–7 February 2025. Stellenbosch: Department of Forest and Wood Science, Faculty of AgriSciences, Stellenbosch University, 72.

Pajares, G., Peruzzi, A. & Gonzalez-de-Santos, P. 2013. *Sensors in Agriculture and Forestry*. Basel: MDPI.

Peerbhay, K. & Maresch, M. 2025. Adapting computer vision technologies for measuring plant quality and vitality at Sappi's largest commercial nursery. *In:* Erasmus, J., Talbot, B., Dlokweni, P. & Grube, G. (eds.) *Precision Forestry: At the Dawn of Artificial Intelligence (AI)*, 3–7 February 2025. Stellenbosch: Department of Forest and Wood Science, Faculty of AgriSciences, Stellenbosch University, 72.

Pérez-Carrasco, M., Karelovic, B., Molina, R., Saavedra, R., Cerulo, P. & Cabrera-Vives, G. 2022. Precision silviculture: Use of UAVs and comparison of deep learning models for the identification and segmentation of tree crowns in pine crops. *International Journal of Digital Earth,* 15, 2223–2238.

Ponsse. 2024. *Solutions that Improve the Operator's Ergonomy and Make Working Easier* [Online]. Finland: Ponsse. Available: https://news.ponsse.com/en/solutions-that-improve-the-operators-ergonomy-and-make-working-easier/ [Accessed 2 May 2025].

Pöry. 2014. *Reinventing Plantation Forestry*. Brazil: Pöry. https://silo.tips/download/reinventing-plantation-forestry.

Poschen, P. 1993. Forestry, a safe and healthy profession? *International Journal of Forestry and Forest Industries,* Unasylva - No. 172, 44. Available: https://www.fao.org/4/u8520e/u8520e03.htm#forestry,%20a%20safe%20and%20healthy%20profession

Pozdnyakova, U. A., Golikov, V. V., Peters, I. A. & Morozova, I. A. 2019. Genesis of the revolutionary transition to industry 4.0 in the 21st century and overview of previous industrial revolutions. In Popkova, E. G. & Sergi, B. S. (eds.) *Industry 4.0: Industrial Revolution of the 21st Century.* Cham: Springer, 11–19. https://doi.org/10.1007/978-3-319-94310-7_2.

Qi, H. & Liang, X. 2024. Automated first-order tree branch modeling at plot-and individual-tree-levels from close-range sensing for silviculture and forest ecology. *IEEE Transactions on Geoscience and Remote Sensing*, 62, 1–21.

Ramantswana, M. 2022. *Is There Room for Electric Tractors in Forestry?* [Online]. www.forsilvitech.com: Forsilvitech. Available: https://www.forsilvitech.com/article/is-there-room-for-electric-tractors-in-forestry/ [Accessed 2 May 2025].

Ramantswana, M. 2023a. *Ecoforst's T-Winch Enables Mechanised Regeneration on Steep Slopes* [Online]. www.forsilvitech.com: Forsilvitech. Available: https://www.forsilvitech.com/article/ecoforst-s-t-winch-enables-mechanised-regeneration-on-steep-slopes/ [Accessed 1 May 2025].

Ramantswana, M. 2023b. *Steep Terrain Solutions for Mechanised Ground-based Silviculture Equipment* [Online]. www.forsilvitech.com: Forsilvitech. Available: https://www.forsilvitech.com/article/steep-terrain-solutions-for-mechanised-ground-based-silviculture-equipment/ [Accessed 1 May 2025].

Ramantswana, M. 2023c. *Using the TimberMAX Traction-Winch Assist in Land Clearing Operations* [Online]. www.forsilvitech.com: Forsilvitech. Available: https://www.forsilvitech.com/article/using-the-timbermax-traction-winch-assist-in-land-clearing-operations/ [Accessed 1 May 2025].

Ramantswana, M., Brink, M., Little, K., Spinelli, R. & Chirwa, P. 2021a. Perspectives and drivers of modernisation of silviculture re-establishment in South Africa. *Southern Forests: A Journal of Forest Science,* 83, 79–87.

Ramantswana, M., Spinelli, R., Brink, M. & Little, K. 2021b. A forecast of future silviculture re-establishment technologies in plantation forestry. *Scientia Forestalis,* 49, e3488.

Richardson, B. 2024. Are uncrewed aerial spraying systems the future for forestry pesticide application? *New Zealand Journal of Forestry Science,* 54.

Robinson, D. G., Trites, D. G. & Banister, E. W. 1993. Physiological effects of work stress and pesticide exposure in tree planting by British Columbia silviculture workers. *Ergonomics,* 36, 951–961.

Rubilar, R. A., LEE Allen, H., Fox, T. R., Cook, R. L., Albaugh, T. J. & Campoe, O. C. 2018. Advances in silviculture of intensively managed plantations. *Current Forestry Reports,* 4, 23–34.

Sandino, J., Pegg, G., Gonzalez, F. & Smith, G. 2018. Aerial mapping of forests affected by pathogens using UAVs, hyperspectral sensors, and artificial intelligence. *Sensors,* 18, 944.

SCA. 2024. *Real operation for remote-controlled soil scarification machine from SCA* [Online]. Sundsvall, Sweden: SCA. Available: https://www.sca.com/en/media/news/2024/real-operation-for-remote-controlled-soil-scarification-machine-from-sca/ [Accessed 3 May 2025].

Schwegman, K., Spinelli, R., Magagnotti, N., Ramantswana, M. & Mcewan, A. 2023. A comparison of excavator-based harvester productivity among different machine operators in a clear-cut Eucalyptus plantation forest, South Africa. *New Zealand Journal of Forestry Science,* 53.

Seenivasan, R. & Parthiban, K. 2017. Precision silvicultural technology for industrial plantations. *In:* Parthiban, K. T. & Seenivasan, R. (eds.) *Forestry Technologies-A Complete Value Chain Approach.* Jodhpur: Scientific Publishers, 194–216. ISBN 9789386102607.

Sharma, A. & Singh, B. J. 2020. Evolution of industrial revolutions: A review. *International Journal of Innovative Technology and Exploring Engineering,* 9, 66–73.

Shivaprakash, K. N., Swami, N., Mysorekar, S., Arora, R., Gangadharan, A., Vohra, K., Jadeyegowda, M. & Kiesecker, J. M. 2022. Potential for artificial intelligence (AI) and machine learning (ML) applications in biodiversity conservation, managing forests, and related services in India. *Sustainability,* 14, 7154.

Silversides, C. 1984. Mechanized forestry, World War II to the present. *The Forestry Chronicle,* 60, 231–235.

Singh, D., Arya, S., Gupta, B., Kaushik, D., Arya, V., Kumar, U. & Singh, K. 2021. Applications of nanotechnology in forest management. *Journal of Nanoscience and Nanotechnology,* 21, 3466–3480.

Skilton, M. & Hovsepian, F. 2018. *The 4th Industrial Revolution: Responding to the Impact of Artificial Intelligence on Business.* Cham: Springer. https://doi.org/10.1007/978-3-319-62461-7.

Skinner, C. N. 2007. Silviculture and forest management under a rapidly changing climate. *In:* Powers, R. F., (ed.) *Restoring Fire-adapted Ecosystems: Proceedings of the 2005 National Silviculture Workshop,* 6–10 June 2005, Tahoe City. USDA Forest Service, 21–32.

Smith, D. 1986. *The Practice of Silviculture*, 8th ed. New York: John Wiley.

Sobocki, S., Adamczyk, F., Tylek, P., Szaroleta, M., Ż Urowski, K., Kormanek, M., Szewczyk, G., Tabor, S., Zwierzyński, M. & Kaź Mierczak, M. 2023. A seedling collection unit of a mobile automatic device for forest tree planting—An extended operating concept. *Forests,* 14, 2420.

Spencer, G. & Torres, P. M. 2022. New can bus communication modules for digitizing forest machines functionalities in the context of forestry 4.0. *IEEE Access,* 11, 9058–9066.

Sullman, M. J. & Byers, J. 2000. An ergonomic assessment of manual planting Pinus radiata seedlings. *Journal of Forest Engineering,* 11, 53–62.

Szmyt, J. & Dering, M. 2024. Adaptive silviculture and climate change—A forced marriage of the 21st century? *Sustainability,* 16, 2703.

Tigercat. 2022. Canada: TIgercat. Available: https://www.tigercat.com/féature/turnaround/ [Accessed 2 May 2025].

Trites, D. G. 1992. *Ergonomics of tree planting work among British Columbia forest workers*. Master's thesis, Simon Fraser University, Burnaby, BC, Canada.

Troxler, P. 2013. Making the 3rd industrial revolution. *In:* Walter-Herrmann J. & Büching, C. (eds.) Fab Labs*: Of Machines, Makers and Inventors.* Bielefeld: Transcript Publishers, 1–16.

Van Goor, C. 1978. Trends in forestry research and its implication on silviculture. *Irish Forestry,* 34 (2), 62–76. Available: https://journal.societyofirishforesters.ie/index.php/forestry/article/view/9379.

Vergouw, B., Nagel, H., Bondt, G. & Custers, B. 2016. Drone technology: Types, payloads, applications, frequency spectrum issues and future developments. *In:* Custers, B. (ed.) *The Future of Drone Use: Opportunities and Threats from Ethical and Legal Perspectives.* The Hague: T.M.C. Asser Press, 21–45. https://doi.org/10.1007/978-94-6265-132-6_2.

Visser, R. & Berkett, H. 2015. Effect of terrain steepness on machine slope when harvesting. *International Journal of Forest Engineering,* 26, 1–9.

Visser, R. & Obi, O. F. 2021. Automation and robotics in forest harvesting operations: Identifying near-term opportunities. *Croatian Journal of Forest Engineering: Journal for Theory and Application of Forestry Engineering,* 42, 13–24.

Visser, R. & Stampfer, K. 2015. Expanding ground-based harvesting onto steep terrain: A review. *Croatian Journal of Forest Engineering: Journal for Theory and Application of Forestry Engineering,* 36, 321–331.

Wang, T., Zuo, Y., Manda, T., Hwarari, D. & Yang, L. 2025. Harnessing artificial intelligence, machine learning and deep learning for sustainable forestry management and conservation: Transformative potential and future perspectives. *Plants,* 14, 998.

Wu, Z., Chen, Y., Zhao, B., Kang, X. & Ding, Y. 2021. Review of weed detection methods based on computer vision. *Sensors,* 21, 3647.

Xu, M., David, J. M. & Kim, S. H. 2018. The fourth industrial revolution: Opportunities and challenges. *International Journal of Financial Research,* 9, 90–95.

Ziatdinov, R., Atteraya, M. S. & Nabiyev, R. 2024. The fifth industrial revolution as a transformative step towards society 5.0. *Societies,* 14, 19.

Zou, W., Jing, W., Chen, G., Lu, Y. & Song, H. 2019. A survey of big data analytics for smart forestry. *Ieee Access,* 7, 46621–46636.

6 Relationship between silviculture and harvesting

Muedanyi M. Ramantswana and Saulo P. S. Guerra

6.1 Introduction

This chapter will explore the intricate relationship between silviculture and harvesting operations in forestry. It will focus on how establishment and tending activities can impact harvesting, and conversely, how harvesting can affect silvicultural practices. The discussion will primarily centre on even-aged clear-fell silvicultural systems, as commonly employed in South Africa and Brazil, which involve regeneration, tending, and harvesting. This chapter will also address sustainability within this context and how effective planning will facilitate optimised value chains. In this chapter, the reader will learn about various activities in harvesting and silviculture, their impacts, and how their effects can be mitigated or reduced. Ultimately, it will propose improved efficiencies and waste reduction in both silviculture and harvesting. This information will be presented to enable various stakeholders in forestry to think holistically about actions and changes, particularly in anticipation of adverse effects that may not be immediately apparent but that will be experienced later in the rotation cycle.

6.2 Sustainability in forestry

The sustained use of forest resources for current and future generations has been a core theme in forestry. Essentially, the core theme revolves around the principle of continuity in the long-term use of resources through responsible management practices that balance the preservation and utilisation of forest resources (Sarkar, 1998). According to Schweier et al. (2019), sustainable forest management focuses on the current and future use of forests, which maintains their biodiversity, productivity, regeneration capacity, and growth to meet ecological, social, and economic objectives. The complexity is in managing the forest's regenerative ability in a manner that benefits current needs without compromising future needs (MacDicken et al., 2015).

Forest management influences the overall productivity of the site through the regeneration of the correct species suited to the site, executing silvicultural practices that facilitate optimal growth and improve site quality (e.g.

DOI: 10.4324/9781003391227-6

fertilisation, soil amelioration) (Burger, 1996). Evans (1999) made the following conclusions after assessing the long-term productivity of plantation forests:

i A quantification of the yield from successive rotations of trees showed there is no significant trend that plantation forestry is unsustainable. Reductions in yield can be attributed to poor silvicultural practices
ii Evidence from different countries indicates that the current growth rates have improved due to changes in the environment and improvements in silviculture
iii Silviculture interventions, such as genetic material improvement, suggested the potential for significant and sustained gains over several rotations

Even though plantation forests are generally sustainable, there are external factors which will influence their sustainability. Risks and hazards, such as climate change, increase the vulnerability of forests to hazards like extreme weather events, fires, and windstorms (Pawson et al., 2013). Furthermore, pests and pathogens may have a greater impact on stand survival and growth (Wingfield et al., 2008). Therefore, adapting silviculture to expected conditions will become crucial to ensure that plantation forests are sustained in the long term (Freer-Smith et al., 2019).

To improve sustainable forest management, forest certification was initiated in the early 1990s (Perera and Vlosky, 2006). Certification systems promote sound forest management by facilitating access to markets or better product prices for wood produced from forests with evidence of responsible management (Siry et al., 2005). Certification aims to assure consumers about the quality of forest management executed in forestry sources and the value chain. Organisations such as the Forest Stewardship Council (FSC) and the Programme for the Endorsement of Forest Certification (PEFC), among many other certification bodies, are involved in certification by using principles, criteria, and indicators to measure responsible forest management practices (Rametsteiner and Simula, 2003). The execution of silviculture and harvesting activities is comprehensively assessed to ensure that all activities are conducted in ways that facilitate social benefits, are environmentally appropriate, and are economically viable (van der Ven and Cashore, 2018). South Africa's total FSC-certified forestry area is 1.4 million hectares (Oberholzer, 2021). In Brazil, the total certified area is 10.6 million hectares, and the total planted area is 7.8 million hectares. The FSC-certified forestry area is 3.0 million hectares, the PEFC-certified forestry area is 2.5 million hectares, and the FSC/PEFC (double certification) covers 2.3 million hectares (Brazilian Tree Industry, 2024).

6.3 Planning of operations

Silviculture tasks involve a series of scheduled activities (site preparation, spacing, fertilisation, control of leaf-cutting ants and weeds, irrigation, pruning, and thinning) necessary to achieve optimal yield from a forest stand (Mendonça et al., 2022). These activities must be carried out correctly and promptly to prevent unnecessary costs. For instance, a delay in chemical weeding after planting can reduce tree growth due to heightened competition and increase weeding expenses, as more costly methods (e.g. manual slashing) may be needed to control the weeds (Little et al., 2005). In plantation forestry, the landowner must practice good silviculture to produce consistent yields, ensuring the costs of ownership and management are covered (Nyland, 2016). Well-managed plantation forests improve stand productivity and enable increased wood production from limited areas (Siry et al., 2005).

Silviculture operations in South Africa remain largely manual. As machines and various technologies are increasingly used for silviculture activities, planning becomes more complex because machines are more expensive and less adaptable than manual teams (Ramantswana et al., 2021). In Brazil, the overall level of mechanisation for silvicultural operations with eucalyptus is mechanised, while manual practices are used for pine plantations, according to Guerra et al. (2025).

Key challenges for mechanised silviculture include accessing steep terrain, dealing with physical obstacles, and weighing the cost–benefit comparison to manual operations (Baker, 2018). Nonetheless, improved planning techniques – through strategic, tactical, and operational planning – help ensure that work and machine usage are well balanced to achieve optimal utilisation, reduce costs, and increase productivity.

Proper planning helps to reduce the negative impacts of silviculture and harvesting activities. Planning enables foresters to anticipate the potential effects on the environment, as well as the social and economic components of the forestry plantation. Planning in the long, medium, and short term enables forestry organisations to identify risks and opportunities to improve their goals. Inadequate planning can lead to poor environmental, social, and economic impacts, which affect the overall sustainability of the forestry enterprise. Adhering to the continuous improvement cycle (planning, doing, checking, and acting – PDCA) can steer forest organisations in the right direction when it comes to managing risks that may arise from safety, productivity, environmental, quality, and social aspects (Brink, 2012). The following section examines the impacts of harvesting on silviculture (and vice versa) and interventions that can be applied to mitigate or reduce these impacts. The impacts are not exhaustive because some impacts have not been studied, or the extent of their effect is not well understood. This section aims to create an awareness of the interconnectedness and dependency between silviculture and harvesting. As mentioned in the introduction (Chapter 1), even though silviculture refers to regeneration, stand tending, and harvesting, we have practically categorised the processes into silviculture (mainly referring to regeneration and tending activities) and (ii) harvesting (clear-felling, extraction, and transport). Hence, within the context of this book and this chapter, the term silviculture is used with a focus on the regeneration and tending process stages. In this chapter, the key relationships between "silviculture and harvesting" are identified to identify the interdependencies and impacts of the silviculture stages (regeneration and tending) on harvesting stages and vice versa.

6.4 Harvesting effects on modern silviculture

Harvesting operations significantly impact the way silviculture activities are conducted (Figure 6.1). Poor harvesting practices can increase costs in silviculture operations. Therefore, understanding how different harvesting activities can potentially affect silviculture practices is essential (Schönau, 1989).

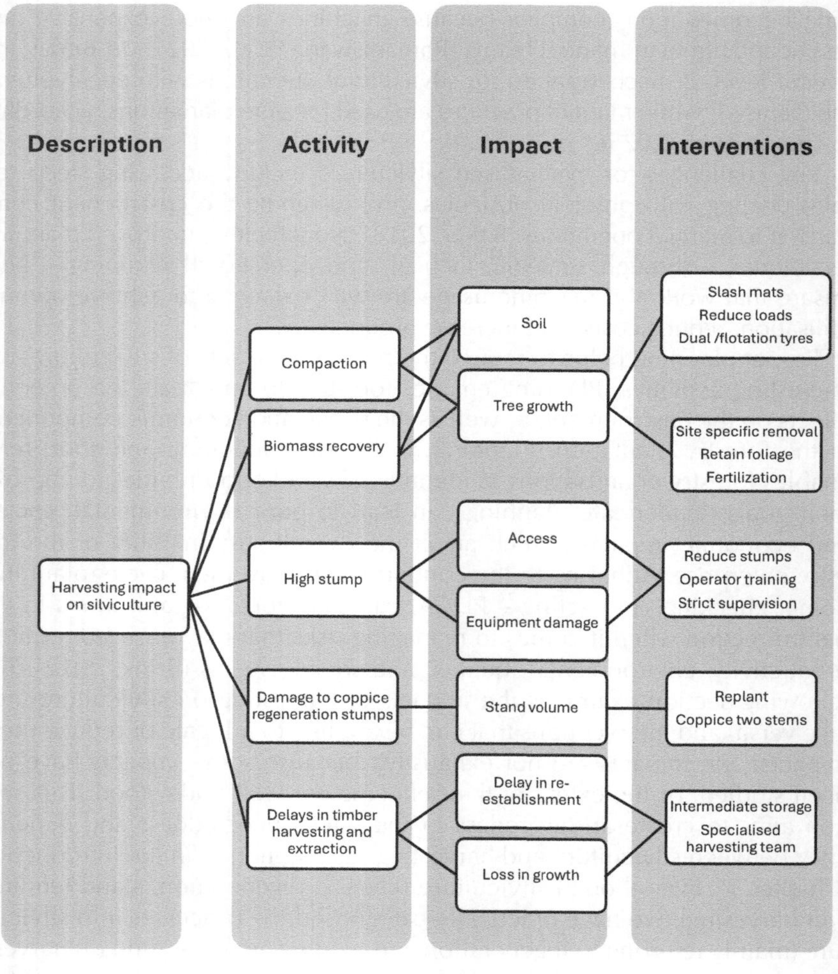

Figure 6.1 Summary of harvesting impacts on silviculture

Failing to comprehend the impacts of harvesting practices on silviculture and long-term sustainability can lead to:

- Increased silviculture costs
- Low survival
- Reduced growth
- Ununiform stand
- Reduced yields and increased harvesting costs

6.4.1 Site impacts (compaction, erosion, rutting)

Forest soil is essential because it provides a surface for tree growth, a habitat for flora and fauna, and a resource for nutrient recycling, regulates water courses, and acts as a buffer for atmospheric chemicals (Worrell and Hampson, 1997). Soil comprises inorganic material from the parent bedrock and organic material derived from vegetation (Marshall, 2000). An understanding of the soil's chemical and physical properties, along with the effects of management practices on soil quality, enables better management of potential impacts. Soil productivity can be enhanced through various practices, including drainage, tillage, and the addition of materials such as lime and fertiliser. However, soil can also be negatively affected by erosion, loss of organic matter, and damage to physical structure (Schoenholtz et al., 2000).

6.4.1.1 Description of impact

Forest operations can cause significant soil disturbance if not conducted properly. The extent of impact depends on changes in soil conditions and the proportion of affected areas. When harvesting activities occur during wet weather, it results in compaction that harms the topsoil structure and water infiltration (Grey and Jacobs, 1987). Soils high in clay and silt are more susceptible to severe compaction as moisture levels increase (Smith et al., 1997). The soil can be negatively affected through compaction, rutting, erosion, and pollution, which can alter its chemical, physical, and biological properties (Marchi et al., 2018). Vehicular traffic from machinery, such as harvesters and forwarders, can cause compaction at various levels. As the machine with its weight moves across the soil surface, it directly increases soil bulk density beneath and alongside the tracks (Dias Junior et al., 2007, Labelle and Jaeger, 2011).

Soil compaction can decrease water infiltration and increase runoff and erosion on slopes. Furthermore, compaction reduces microorganism activity and inhibits the roots from accessing the water, oxygen, and nutrients the trees need to grow (Picchio et al., 2020). Compaction can lead to a reduction in the overall productivity of the site (Marshall, 2000).

6.4.1.2 Description of measures to mitigate or reduce the impacts

The effects of soil compaction vary depending on specific sites. At certain sites (sandy soils), a low to medium level of soil compaction has been found to enhance tree growth in soils with low bulk density by improving water-holding capacities (Smith, 2000, Smith and du Toit, 2005). Compaction can be significantly reduced by not harvesting when the soil is wet and by reducing the number of extraction routes where possible. Operationally, erosion of soil on the harvested site can be reduced by considering the harvesting extraction method, erodibility of the soil, the slope of the area, and the rainy period (Grey and Jacobs, 1987).

Slash mats (medium to heavy) are an effective way to reduce compaction on the site (Han et al., 2009, Rietz, 2010). However, it is essential to understand that as the number of passes increases, the effectiveness of the slash mat also decreases as the material begins to break down. Adaptations can also be made by reducing the load sizes and using high-flotation tyres. When comparing the cut-to-length (CTL) and full-tree methods, the CTL method causes less overall compaction and requires less area for wood extraction. Due to most compaction occurring during the first few passes, extraction must be limited to the same extraction routes used in previous rotations if possible to avoid damaging more areas (Han et al., 2009). Furthermore, sites that have been compacted can be remediated mechanically by tilling or ripping to loosen the soil (Ares et al., 2005, Greacen and Sands, 1980).

6.4.2 Biomass recovery

Residues are biomass materials generated as trees grow in the forest. Once primary products are recovered, residues or slash typically remain on the site. Woody biomass refers to harvesting slash, which can comprise tops, bark, branches, offcuts, and needles or leaves (Titus et al., 2021), as well as stumps and roots (Guerra and Eufrade-Junior, 2019). This material is considered of lower value compared to the conventional primary products such as pulpwood and sawtimber. The quantity and form in which the residues are presented on a particular site depend on the harvesting method used for that area. The CTL method broadcasts residues, while the tops and branches are left in the infield. In the tree length method, the tops and branches are retained on the site, and only the bole is extracted to the roadside. In the full-tree method, trees with their tops and branches are extracted, and therefore, residues accumulate on the roadside (Langin et al., 2010, Udali et al., 2025). In South Africa, biomass harvesting is conducted to increase income, reduce fuel loads, optimise resource utilisation, and meet the need for renewable energy. Consequently, the demand for biomass material has increased (Petrie, 2014). Biomass or residues that would typically be left on the site are recovered based on the landowner's objectives (Rawat et al., 2024). In Brazil, similar to South Africa, the harvesting of biomass, particularly forest residues, and harvesting below-ground eucalyptus biomass, such as stumps and roots, offers a strategic opportunity to significantly boost income for forest owners and optimise raw material supply for industrial uses like energy generation. This practice is increasing and becoming economically feasible due to high energy input costs and directly supports Brazil's renewable energy targets, helping to create a cleaner energy matrix, while also aiding site preparation and renewal for future planting cycles (Guerra and Eufrade-Junior, 2019, de Souza Kulmann et al., 2022).

In Brazil, a significant focus of recovery efforts has been on stumps and roots, as they represent a considerable portion of the total tree biomass. For instance, studies on *Eucalyptus* plantations indicate that stumps and roots can account for 27.9% of the total dry biomass of an average tree, or even up to 20%–41% of the total tree biomass in some cases (Guerra and Eufrade-Junior, 2019). The increasing interest in biomass recovery from multiple objectives, such as increasing forest landowners' revenue, reducing fuel loads, optimising resource utilisation, meeting renewable energy needs, and facilitating silviculture operations. Removal of stumps and roots significantly improves site access, facilitating silvicultural manual, semi-mechanised, and mechanised operations like soil preparation, re-alignment of planting line, and new standing tending (Walmsley and Godbold, 2010) and avoiding a loss of functional area for new plantings or subsequent silvicultural operations (Pereira da Silva Correia et al., 2020).

6.4.2.1 Description of impact

When biomass is retained on the site, it serves as soil cover, which can cushion the soil from impacts such as compaction by harvesting and extraction equipment. The residues also serve as a source of nutrients for the site if retained and allowed to decompose. When biomass material (litter layer and harvesting residues like bark, branches and foliage) is removed due to biomass collection, the nutritional composition of the site can be altered, which leads to lower tree growth (du Toit et al., 2004, du Toit, 2008, Walmsley et al., 2009). The concentrations of N and P, as well as soil organic carbon, can be reduced when residues are removed (Rietz, 2010). Furthermore, the removal of biomass and harvest residues can lead to the loss of soil organic carbon, increased erosion and increased impact of machinery on the soil surface (Titshall et al., 2013). The extent of the effect of biomass removal on tree growth in future rotations will depend on the site, removal intensity, and silvicultural interventions (e.g. fertilisation) (Achat et al., 2015).

Residues retained on site can also impact site preparation accessibility (Nieto Lawrence, 2022). The execution of operations like pitting and planting can be hampered by residues on the site, which may influence the survival of trees and influence how operations like pitting and planting are executed (Ndlovu et al., 2019). For example, when pitting, the operator needs to clear the spot before pitting the soil, which can reduce the machine's productivity (Jacob, 2022).

6.4.2.2 Description of measures to mitigate or reduce the impacts

The recovery of biomass needs to be planned to assess the risks associated with the removal of residues on various sites (Achat et al., 2015). Assessments should consider:

- The biomass material which can be collected and the amount of nutrients that will be taken out
- The harvesting method to be used to harvest and recover biomass
- Collection and processing methods to be applied
- Biomass recovery should not be conducted on sensitive sites

To improve site nutrition, harvesting methods must facilitate needle drop instead of harvesting the entire tree and processing it into biomass. It is recommended that biomass should only be recovered from robust sites. Fertilisation of nutrient-poor sites can improve tree growth (du Toit et al., 2001). The addition of fertiliser to sites where residues are removed can assist with stand growth and overall productivity gains. However, when residues and forest floor material are removed, significant fertilisation is required to remediate the nutrient losses (Garrett et al., 2021). However, the removal of residual biomass, especially below-ground biomass, offers notable advantages for forest operations, management, and bioenergy production; it demands careful consideration of its potential impacts on soil nutrient sustainability and overall forest production costs. Strategically, maintaining an aboveground nutrient-rich compartment on the site is recommended for forest long-term site quality, making the harvest of stumps and roots potentially less detrimental to the removal of other tree fractions (leaves, branches, bark) (de Souza Kulmann et al., 2022).

6.4.3 High stumps

When trees are felled, the stump remains in the ground. Stumps are typically about 10 cm high or less. The height of stumps can be influenced by the felling equipment (chainsaw, harvester, or feller buncher), the trees (planted or coppiced), terrain conditions (steep slopes) and the underfoot conditions (vegetation or litter layer) (Ramantswana et al., 2017, Han and Renzie, 2005, Hall and Han, 2006).

6.4.3.1 Description of impact

When stumps are high, wood that could potentially be sent to the processing plant is wasted in the infield. Furthermore, stumps impede access to the site for silviculture activities using machines. Silviculture equipment and attachments (e.g. pitting machines and tractor-towed planters) used for re-establishment often lack adequate clearance to drive over high stumps. High stumps can cause punctures and damage the undercarriage of machines.

6.4.3.2 *Interventions to mitigate the impact*

High stumps that can be reduced or removed to facilitate access to the site. High stumps can be reduced by using mulchers or grinders to break down the stump wood into finer pieces and distribute the material throughout the stand (Zitha et al., 2023). High stumps can be reduced by cutting the lower part with a chainsaw. This operation is slow and risky for the operator to perform. A machine like a stump clipper can be used to cut the stumps at ground level. The stumps can be left in the infield to decompose or recovered as biomass material. Stumps can be harvested by being uprooted from the ground and recovered for energy purposes (Laitila et al., 2008, Berg et al., 2015). When the stumps are removed, access to the site is improved.

6.4.4 *Coppice regeneration capacity*

Coppice regeneration involves the ability of the stump to resprout from the epicormic buds or lignotubers, which are located in the live bark and cambium on the tree stump after felling (Little et al., 2002). In plantation forests, eucalypts are known to have a high resprouting capability (Ferraz Filho et al., 2014). Coppicing can be applied to reduce regeneration costs and shorten regeneration time (Silva et al., 2020) (Refer to Chapter 3.7) for details on coppicing.

6.4.4.1 *Description of the impact*

The extent to which stumps are damaged during harvesting can influence the success of a coppice regeneration (Ferraz Filho et al., 2014). As harvesting machines operate, they tend to damage the sprout-producing buds (Hakamada et al., 2022). If machines damage the stumps during harvesting, the coppice resprout is lower, and the stocking may be insufficient to regenerate the stand successfully (Crous and Burger, 2015). The harvesting system used can have a direct impact on the survival and uniformity of a stand. Damage can be caused by vehicle movement, tear-outs during felling and stripping (Schwegman et al., 2018). Hakamada et al. (2022) found that using a feller buncher yielded lower coppice survival and uniformity compared to using a chainsaw for felling. Generally, the application of mechanised felling can inflict damage on stumps, which is a problem because the viability of the next rotation is dependent on a high percentage of coppice sprouting (Spinelli et al., 2017). The coppice regeneration capacity of a forest stand needs to be maintained or increased for sustainable forest operations (Marchi et al., 2018).

6.4.4.2 *Interventions to mitigate the impact*

Coppice regeneration can be improved by reducing harvesting damage on the stumps. Operators need to be informed and trained about the planned regeneration methods so that they can avoid damaging the stumps, which are an essential resource for coppice sprouting. Harvesting methods which cause the least amount of damage to stumps can also be considered.

6.4.5 Delays in timber harvesting and extraction

Timber harvesting of stands is planned and carried out according to the scheduled annual operational plan. However, due to various factors, some portions of the stands may not be harvested and extracted because they cannot be accessed by the main harvesting system. Additionally, timber that might have been felled and processed could be held in the field for weeks or months until it meets market standards.

6.4.5.1 *Description of impact*

The challenge with delayed harvesting and extraction of timber from a site is that it restricts the ability to execute silviculture regeneration activities. For example, if a difficult pocket is left standing in a harvested stand, regeneration activities cannot be conducted around the pocket, as it will become landlocked and inaccessible. Attempts to harvest and extract the stand after regeneration will result in damage to young trees and unnecessary losses.

When timber is harvested and cannot be removed from the site until it meets market specifications (e.g. moisture requirements), it increases the temporary unplanted area. Even though over 90% of the harvested stand may be accessible for regeneration, the activities cannot be executed efficiently. For example, in South Africa, if a stand has been harvested, timber is extracted and stacked on the roadside, waiting to meet market requirements. If harvest residues must be burnt, there is a risk that the fire may burn the timber on the roadside; hence, the stand is left untreated until all wood is transported. Furthermore, if regeneration activities are conducted on the available land, the small portion covered by timber also needs to be regenerated once the wood is transported. This can increase regeneration costs (e.g. machine relocation costs) and decrease stand uniformity as the trees planted later will be smaller and will not catch up with the rest of the planted stand. Similarly, in Brazil, after harvesting, *Eucalyptus* logs have a high moisture content, ranging from 50% to 60% on a wet basis, so they require drying before industrial processing. Typically, the logs are stored for 90–180 days in uncovered piles exposed to natural air drying, which helps achieve the desired low moisture content – below 35% on a wet basis – to boost fuelwood's energy value for combustion and meet the operational and logistical demands of large forestry companies (Eufrade-Junior et al., 2021).

6.4.5.2 *Interventions to mitigate the impact*

To reduce the impact of timber harvesting and extraction delays on silviculture, careful planning is essential to match the appropriate harvesting systems to site conditions, thereby preventing difficult pockets from being left behind. Another option could be to have a smaller, specialised clean-up team that handles the challenging pockets immediately after the main harvesting is completed, avoiding delays in regeneration activities. When timber must be kept on site due to market demands, it should be stored at a temporary location, such as an intermediate storage area, to prevent loss of stand growth time caused by storing timber on the harvested site.

6.5 Silviculture operations' effect on harvesting operations

Silviculture activities conducted during the regeneration and stand-tending phases can have a direct impact on wood quality (Zobel, 1992), as well as the productivity and efficiency of harvesting operations at the rotation end when clear-felling is conducted (Ackerman and Rietz, 2014). When considering silvicultural treatments, type 2 responses characterised by a sustained change in stand productivity (volume increment) are important (Snowdon, 2002). The growth responses resulting from treating a particular stand's limiting factor can have a long-term effect on the stand's growth. Forest managers need to base their decisions on objective information about the long-term effects of silvicultural treatments (Snowdon, 2002). The impact of silviculture activities or methods needs to be assessed to avoid optimising one activity now to the detriment of another in a few years.

The main factors that will have a direct influence on harvesting can be manipulated from a silvicultural perspective (Figure 6.2).

6.5.1 Execution of regeneration activities

The execution of regeneration activities, such as residue management and planting, can impact the survival, growth, and uniformity of the stand.

6.5.1.1 Description of impact

Site preparation activities must be conducted in a manner that achieves high survival, optimal tree growth, and uniformity. For example, low-intensity slash burning should be practised (De Ronde et al., 1990); the planting positions must be adequately prepared, and the planting operation must be executed correctly, avoiding J rooting and applying the required irrigation if applicable on the site (Hechter, 2019). Poor silvicultural practices can lead to high mortalities and poor tree growth. Sites that require improvement (e.g. compacted sites) must be assessed and actions taken (e.g. ripping compacted sites) to correct tree growth-restricting features prevalent on the site (du Toit et al., 2010).

6.5.1.2 Interventions to mitigate the impact

The effects of poor silvicultural practices can be mitigated through the training of staff involved in executing the activities. Sometimes, due to labour

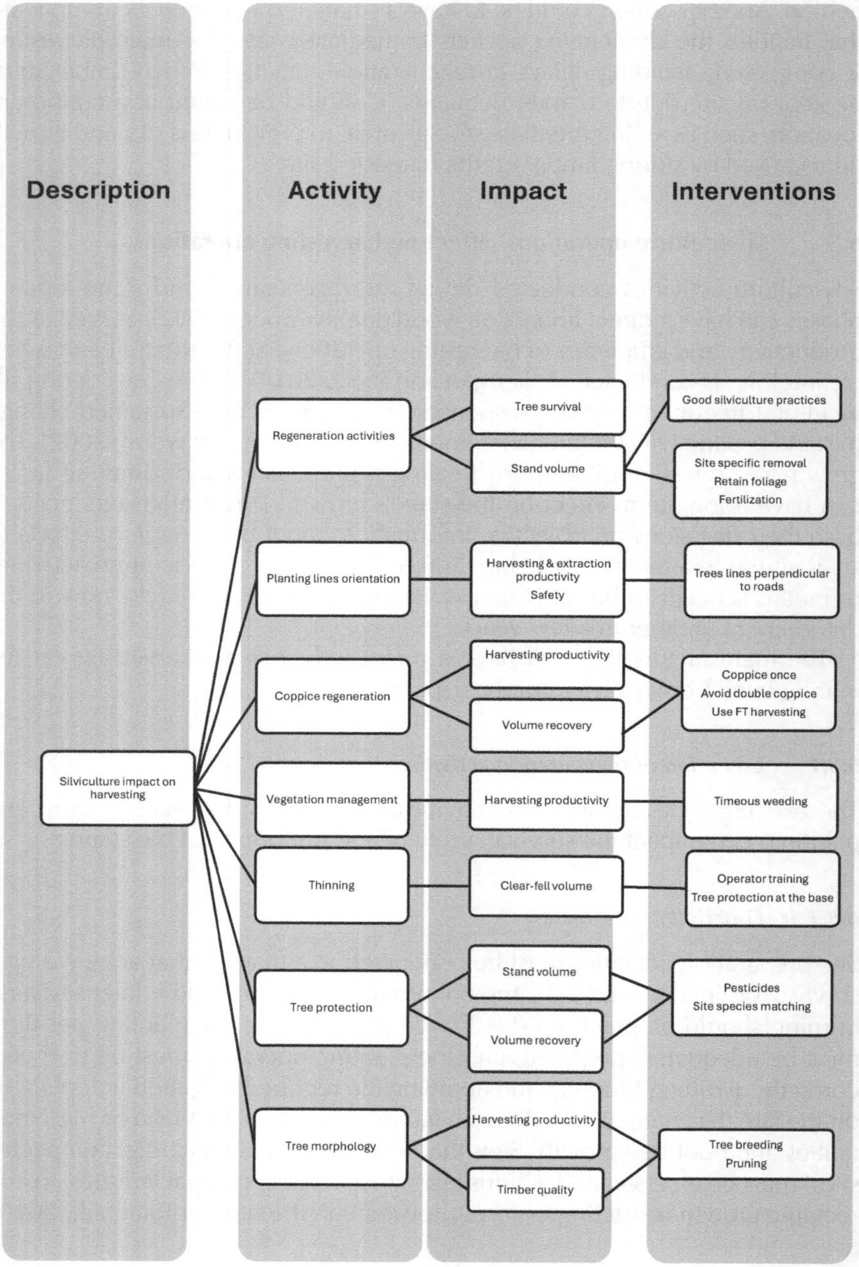

Figure 6.2 Silviculture impacts on harvesting

turnover, contractor changes, and management changes, fundamental knowledge can be lost; therefore, refresher training is also important to ensure the core principles of silviculture are adhered to. Supervision of operations by the forester and team supervisor is also essential for identifying incorrect practices and promptly applying remedial actions.

6.5.2 *Planting line orientation*

Planting patterns refer to the inter- and intrarow spacing as well as the orientation of lines when a compartment is planted.

6.5.2.1 Description of impact

The configuration of planting lines or tree rows on steep areas can significantly influence the efficiency and safety of harvesting operations (Ackerman and Rietz, 2014). On moderately steep slopes where ripping is used as a cultivation method, it is performed along the contour (Schönau, 1989). Terrain poses the most significant challenge in selecting and applying a harvesting system. Within the terrain, the slope is the most important component because rough and steep terrain causes machines to be unstable and to lose traction. Hence, extraction routes on the hill typically follow a straight up-and-down pattern as much as possible (McEwan et al., 2013), while considering potential environmental impacts (e.g. erosion). If trees are planted across the slope, it poses a challenge for harvesting equipment. The equipment may need to work on side slopes to fell and extract the timber, which poses a high safety risk (instability and overturning).

6.5.2.2 Interventions to mitigate the impact

Proper planning must be conducted to ensure that the soil preparation and planting line orientation will facilitate efficient and safe harvesting. When establishing the compartment, tree lines must be perpendicular to the road to enable better access and facilitate timber harvesting and extraction up and down the slope.

6.5.3 *Coppice regeneration*

For a comprehensive description of coppice regeneration, refer to Section 3.7.

6.5.3.1 Description of impacts

Coppicing can impact the productivity and volume recovered from a stand. Tree volume combined with the composition of stems (stem crowding) in a coppiced stand can significantly decrease the productivity of a harvester compared to harvesting planted (single) stands (Ramantswana et al., 2013,

McEwan et al., 2016). Stem crowding in coppice stands hinders effective felling and head movements (Spinelli et al., 2017). Although coppicing can reduce establishment costs, the increased harvesting costs due to lower productivity, as well as losses due to high stump counts, should be considered.

6.5.3.2 Interventions to mitigate the impact

The effect of coppicing can be reduced by coppicing using an accumulating head for felling, such as a feller buncher, instead of a harvester (Spinelli et al., 2017). Another alternative to consider is ensuring the stand is well stocked for coppice regeneration, which will maximise single stems compared to double or triple stems on the site. Furthermore, coppicing once and then replanting can improve productivity and improve fibre recovery, as lower stumps can be achieved.

6.5.4 Vegetation management

For details about the methods and effects of weeding, refer to Sections 3.4, 3.6 and 4.1.

6.5.4.1 Description of impacts

Tree volume is the most significant driver of productivity in harvesting operations (Langin et al., 2010). Silvicultural factors that reduce or restrict tree volume ultimately affect the volume harvested at the end of the rotation. Vegetation control during the establishment phase is essential, as it can lead to differences in tree volume at the end of the rotation. Little et al. (2018) found that intensively weeded stands yielded 30% more volume in *E smithii* pulpwood when planted in warm temperate regions of South Africa. A lack of weeding can also lead to dense undergrowth in the stand, which can make harvesting trees slow due to poor visibility of the base of the tree and hinder machine movement (Alam et al., 2012, Kärhä, 2006).

6.5.4.2 Interventions to mitigate the impact

Timely weeding operations during the establishment are essential to enhance tree growth and to prevent productivity declines in harvesting machines. To improve harvesting productivity at clear-fell, pre-harvest clearing can be performed (Kärhä, 2006) using clearing saws or tractors with spraying rigs that can fit between the interrows of mature trees. If stands are maintained weed-free, there will be no need for pre-harvest clearing operations.

6.5.5 Tree morphology

Tree morphology refers to the characteristics of a tree, including its stem, branchiness, and canopy structure (Carrillo-López and Yahia, 2019). Each tree has genetic morphological features, some of which can be influenced by the silvicultural treatments (Buckley et al., 1997) to improve the stand productivity and timber quality. Tree breeders have the responsibility of identifying seeds and propagating trees with the characteristics required to meet the landowners' objectives (Van Wyk, 1978). Plants produced from tree breeding programmes must be matched to the growing environment, be disease- and pest-resistant, and deliver the quality of wood demanded by the market (Dvorak, 2012).

6.5.5.1 Description of impacts

Trees with poor morphology make harvesting the final crop more complex. For example, trees with poor form (crooked, forked, multiple leaders, and heavy branches) reduce harvesting productivity, decrease the volume recovered, and affect timber quality (Ramantswana et al., 2013, Labelle et al., 2016, Alam et al., 2012, Van Wyk, 1978). Furthermore, harvesting costs can increase due to lower productivity and higher repair and maintenance costs.

6.5.5.2 Interventions to mitigate the impact

Tree improvement programmes must focus on ensuring the material produced in nurseries and planted in the field meets the landowners' objectives. Continuous improvement and collaboration between tree breeders, silviculturists, and harvesting is vital over the long term (Van Wyk, 1978). Furthermore, with the advent of climate change, where extreme weather events (e.g. strong winds) can be expected, tree species need to be improved to withstand and survive the expected extreme conditions as far as possible (Morris, 2022).

Some of the tree morphology characteristics can be influenced by managing stand density parameters. For example, less dense stands encourage individual tree growth, resulting in more branches and a larger canopy. Pruning of branches can be applied if it adds value to timber and improves harvesting efficiency. Semi-mechanised systems (comprising a motor, manual felling, and processing) may need to be used in cases where mechanised systems are too costly.

6.5.6 Stand thinning

Even though commercial thinning operations are conducted using harvesting equipment, they adhere to silviculture tending prescriptions. Thinning aims to increase the growth of individual trees, which is facilitated by the increased space and reduced competition. The manipulation of the stand through thinning can affect stand density, increase growth, improve wood quality, and affect soil and understorey vegetation (Gonçalves, 2021). For details on thinning, refer to Section 4.4.

6.5.6.1 Description of impacts

When thinning is conducted, it can have an impact on the final volume harvested due to the removal as well as residual stand damage. Residual stand damage (wounds) can occur around the canopy, stem/bole and roots of the trees (Lopes et al., 2018, Dembure et al., 2019). The extent of damage is dependent on the harvesting system applied (Lee et al., 2024, Kizha et al., 2021). The damage can be caused by trees or machines rubbing against the residual trees during felling or extraction. Wounds on residual trees can cause decay, reduce tree growth, increase poor timber quality and make the tree susceptible to diseases and mortality (Tavankar et al., 2020). The response of trees to damage will be dependent on the species and environment.

6.5.6.2 Interventions to mitigate the impact

When conducting thinning, the objective must be to achieve the thinning intensity for the site and age and avoid under- or over-thinning, as it can affect the growth of the stand. Residual tree damage from thinning can be reduced by conducting pre-harvest planning, accurate directional felling, marking of extraction routes, direct supervision, training and using skilled operators and protecting the base of the tree (e.g. tree pads) (Tavankar et al., 2020, Kizha et al., 2021).

6.5.7 Tree protection

From establishment, trees are vulnerable to various mortality-causing biotic (e.g. pests, disease, mammals) and abiotic (e.g. heat, fire, frost, snow) factors (Hechter et al., 2022). Even though the silviculturist may not have control of when these factors affect the stand, practising good silviculture (Hechter et al., 2022), regular stand assessment is important to detect and prevent severe damage to the stand where possible. Tree protection is not the sole responsibility of the silviculturist; the responsibility lies with all interested stakeholders.

6.5.7.1 Description of impacts

When tree stands suffer moderate to high damage or mortalities, it reduces the recoverable volume at clear-fell, which increases the costs of harvesting. In extreme damage cases (e.g. fire damage), salvage operations need to be conducted to recover the wood before the timber quality deteriorates. With the advent of climate change, the occurrence of salvage operations is expected to increase in future (Leverkus et al., 2018).

6.5.7.2 Interventions to mitigate the impact

Various factors, such as site species matching, can reduce the stand's vulnerability to stress-causing factors. Furthermore, the application of pesticides can help protect trees from damage caused by pests and diseases. Regular stand health monitoring is vital to detect stress and mortality in tree stands (Xulu et al., 2019, Peerbhay et al., 2019).

6.6 Summary

This chapter focused on the relationship between silviculture and harvesting operations, specifically the impacts of establishment and tending activities on harvesting operations, as well as the reciprocal effects of harvesting operations on these activities. It noted that South Africa and Brazil employ the even-age clear-fell silvicultural system. This chapter described the importance of sustainability and how planning can facilitate optimised value chains. It explored the impacts of various activities and how their effects could be mitigated or reduced to improve efficiencies and reduce waste in both silviculture and harvesting operations. Furthermore, it highlighted silviculture tasks and their impact on optimal yield, emphasising the need for proper planning to mitigate negative effects on the environment, society, and economy. This chapter also examined harvesting impacts on silviculture, including site impacts like compaction and erosion, biomass recovery, and issues related to high stumps and coppice regeneration capacity. It considered delays in timber harvesting and extraction, as well as their effects on silviculture activities. Finally, this chapter detailed the impacts of silviculture operations on harvesting, covering aspects like regeneration activities, planting line orientation, coppice regeneration, vegetation management, tree morphology, stand thinning, and tree protection, emphasising the interdependence and impacts of both operations on each other.

Review questions

1. How can long-term forest sustainability be improved through better integration of forest value chain activities?
2. Discuss how harvesting operations can have an impact on regeneration operations
3. Discuss how regeneration operations can have an impact on harvesting operations
4. Discuss how stand-tending activities can have an impact on harvesting operations
5. Discuss practical interventions to mitigate the effects described in the questions above

References

Achat, D. L., Deleuze, C., Landmann, G., Pousse, N., Ranger, J. & Augusto, L. 2015. Quantifying consequences of removing harvesting residues on forest soils and tree growth–A meta-analysis. *Forest Ecology and Management,* 348, 124–141.

Ackerman, S. & Rietz, D. 2014. Current operational constraints in the harvesting-silviculture interface: A scoping study. *ICFR Technical Note 02/2014*. Pietermaritzburg: Institute for Commercial Forestry Research.

Alam, M. M., Strandgard, M. N., Brown, M. W. & Fox, J. C. 2012. Improving the productivity of mechanised harvesting systems using remote sensing. *Australian Forestry,* 75, 238–245.

Ares, A., Terry, T. A., Miller, R. E., Anderson, H. W. & Flaming, B. L. 2005. Ground-based forest harvesting effects on soil physical properties and Douglas-Fir growth. *Soil Science Society of America Journal,* 69, 1822–1832.

Baker, M. 2018. *Mechanised Silviculture Opportunities and Challenges for the New Zealand Forest Industry*. New Zealand: Kellog Rural Leadership Programme.

Berg, S. K. N., Nordfjell, T. & Bergström, D. 2015. Effect of stump size and timing of stump harvesting on ground disturbance and root breakage diameter. *Silva Fennica,* 49 (1), Article 1212. https://doi.org/10.14214/sf.1312.

Brazilian Tree Industry. 2024. Ibá 2024 Annual Report. *In:* Tech, E. (ed.). IBÁ.

Brink, M. 2012. Certification as a risk management strategy. *In:* Bredenkamp, B. & Upfold, S. (eds.) *South African Forestry Handbook*. Pretoria: South African Institute of Forestry Research.

Buckley, D. S., Zasada, J. C., Tappeiner, I. & Stone, D. M. 1997. Plant morphological characteristics as a tool in monitoring response to silvicultural activities. *In*: *Communicating the role of silviculture in managing the national forests: Proceedings of the National Silviculture Workshop*. 1997 May 19–22; Warren, PA.: Gen. Tech. Rep. NE-238. Radnor: US Department of Agriculture, Forest Service, Northeastern Forest Experiment Station, 37–41.

Burger, J. A. 1996. *Limitations of Bioassays for Monitoring Forest Soil Productivity: Rationale and Example*. Madison, WI: Wiley Online Library.

Carrillo-López, A. & Yahia, E. M. 2019. Morphology and anatomy. *In:* Yahia, E. M. (ed.) *Postharvest Physiology and Biochemistry of Fruits and Vegetables*. Cambridge, MA: Elsevier. https://doi.org/10.1016/B978-0-12-813278-4.00006-3.

Crous, J. W. & Burger, L. 2015. A comparison of planting and coppice regeneration of Eucalyptus grandis× Eucalyptus urophylla clones in South Africa. *Southern Forests: A Journal of Forest Science,* 77, 277–285.

de Ronde, C., Goldammer, J., Wade, D. & Soares, R. 1990. Prescribed fire in industrial pine plantations. *In:* Goldammer, J. G. (ed.) *Fire in the Tropical Biota: Ecosystem Processes and Global Challenges*. Berlin: Springer, 233–263. https://doi.org/10.1007/978-3-642-75395-4_14.

de Souza Kulmann, M. S., de Jesus Eufrade-Junior, H., Dick, G., Schumacher, M. V., de Azevedo, G. B., Azevedo, G. T. D. O. S. & Guerra, S. P. S. 2022. Belowground biomass harvest influences biomass production, stock, export and nutrient use efficiency of second rotation Eucalyptus plantations. *Biomass and Bioenergy,* 161, 106476.

Dembure, T. P., Mcewan, A., Spinelli, R., Magagnotti, N. & Ramantswana, M. 2019. A comparison between two alternative harvesting systems in the thinning of fast-growing pine plantations under the conditions of low labour cost. *European Journal of Forest Research,* 138, 43–52.

Dias Junior, M. D. S., Fonseca, S. D., Araújo Júnior, C. F. & Silva, A. R. 2007. Soil compaction due to forest harvest operations. *Pesquisa Agropecuária Brasileira,* 42, 257–264.

du Toit, B. 2008. Effects of site management on growth, biomass partitioning and light use efficiency in a young stand of Eucalyptus grandis in South Africa. *Forest Ecology and Management,* 255, 2324–2336.

du Toit, B., Arbuthnot, A., Oscroft, D. & Job, R. A. 2001. The effects of remedial fertilizer treatments on growth and pulp properties of Eucalyptus grandis stands established on infertile soils of the Zululand coastal plain. *Southern African Forestry Journal,* 2001, 9–18.

du Toit, B., Dovey, S., Fuller, G. & Job, R. 2004. Effects of harvesting and site management on nutrient pools and stand growth in a South African eucalypt plantation. *In:* Nambiar, E. K. S., Tiarks, A. T. & Cossalter, J. D. (eds.) *Site Management and Productivity in Tropical Plantation Forests*. Bogor: CIFOR, 31–46.

du Toit, B., Smith, C. W., Little, K. M., Boreham, G. & Pallett, R. N. 2010. Intensive, site-specific silviculture: Manipulating resource availability at establishment for improved stand productivity. A review of South African research. *Forest Ecology and Management,* 259, 1836–1845.

Dvorak, W. 2012. The strategic importance of applied tree conservation programs to the forest industry in South Africa. *Southern Forests: A Journal of Forest Science,* 74, 1–6.

Eufrade-Junior, H. D. J., Spadim, E. R., Rodrigues, S. A., Pai, E. D., Ballarin, A. W. & Guerra, S. P. S. 2021. Impact of rainy and dry seasons on eucalypt fuelwood quality logs stored in piles: A case study in Brazil. *Croatian Journal of Forest Engineering: Journal for Theory and Application of Forestry Engineering,* 42, 291–300.

Evans, J. 1999. Sustainability of forest plantations: A review of evidence and future prospects. *The International Forestry Review,* 153–162.

Ferraz Filho, A. C., Scolforo, J. R. S. & Mola-Yudego, B. 2014. The coppice-with-standards silvicultural system as applied to Eucalyptus plantations—A review. *Journal of Forestry Research,* 25, 237–248.

Freer-Smith, P. H., Muys, B., Bozzano, M., Drössler, L., Farrelly, N., Jactel, H., Korhonen, J., Minotta, G., Nijnik, M. & Orazio, C. 2019. *Plantation Forests in Europe: Challenges and Opportunities*. Joensuu: European Forest Institute Joensuu.

Garrett, L. G., Smaill, S. J., Beets, P. N., Kimberley, M. O. & Clinton, P. W. 2021. Impacts of forest harvest removal and fertiliser additions on end of rotation biomass, carbon and nutrient stocks of Pinus radiata. *Forest Ecology and Management,* 493, 119161.

Gonçalves, A. C. 2021. *Thinning: An Overview. Silviculture*. London: IntechOpen, 41–58.

Greacen, E. L. & Sands, R. 1980. Compaction of forest soils. A review. *Soil Research,* 18, 163–189.

Grey, D. & Jacobs, E. 1987. The impact of harvesting on forest site quality. *South African Forestry Journal,* 140, 60–66.

Guerra, S. & Eufrade-Junior, H. 2019. *Recuperação energética da biomassa de tocos e raízes de florestas plantada*. Botucatu: FEPAF.

Guerra, S. P. S., Oguri, G., Schumacher, M. V., Witschoreck, R. & Balista, B. Z. 2025. *Survey of the Level of Mechanization in Forestry: 2024/2025 Edition*. Piracicaba: IPEF.

Hakamada, R. E., Moreira, G. G., Fernandes, P. G. & Martins, S. D. S. 2022. Legacy of harvesting methods on coppice-rotation Eucalyptus at experimental and operational scales. *Trees, Forests and People,* 9, 100293.

Hall, R. & Han, H.-S. 2006. Improvements in value recovery through low stump heights: Mechanized versus manual felling. *Western Journal of Applied Forestry,* 21, 33–38.

Han, H.-S. & Renzie, C. 2005. Effect of ground slope, stump diameter, and species on stump height for feller-buncher and chainsaw felling. *International Journal of Forest Engineering,* 16, 81–88.

Han, S.-K., Han, H.-S., Page-Dumroese, D. S. & Johnson, L. R. 2009. Soil compaction associated with cut-to-length and whole-tree harvesting of a coniferous forest. *Canadian Journal of Forest Research,* 39, 976–989.

Hechter, U. 2019. *The impact of re-establishment practices on tree survival, growth and uniformity in South African eucalypt plantations.* Masters dissertation, Nelson Mandela University, South Africa.

Hechter, U., Little, K., Chan, J., Crous, J. & DA Costa, D. 2022. Factors affecting eucalypt survival in South African plantation forestry. *Southern Forests: A Journal of Forest Science,* 84, 253–270.

Jacob, Z. 2022. *The effects of slash management methods and pit preparation techniques on pit quality, growth response, operational productivity, and its cost effectiveness at re-establishment.* Master of Science Forestry Dissertation, Nelson Mandela University.

Kärhä, K. 2006. Effect of undergrowth on the harvesting of first-thinning wood. *Forestry Studies,* 45, 101–117.

Kizha, A. R., Nahor, E., Coogen, N., Louis, L. T. & George, A. K. 2021. Residual stand damage under different harvesting methods and mitigation strategies. *Sustainability,* 13, 7641.

Labelle, E. R. & Jaeger, D. 2011. Soil compaction caused by cut-to-length forest operations and possible short-term natural rehabilitation of soil density. *Soil Science Society of America Journal,* 75, 2314–2329.

Labelle, E. R., Soucy, M., Cyr, A. & Pelletier, G. 2016. Effect of tree form on the productivity of a cut-to-length harvester in a hardwood dominated stand. *Croatian Journal of Forest Engineering: Journal for Theory and Application of Forestry Engineering,* 37, 175–183.

Laitila, J., Ranta, T. & Asikainen, A. 2008. Productivity of stump harvesting for fuel. *International Journal of Forest Engineering,* 19, 37–47.

Langin, D., Ackerman, P., Krieg, B., Immelman, A., Potgieter, C., VAN Rooyen, J. & Upfold, S. 2010. *South African Ground based Harvesting Handbook.* Scottsville: Institute for Commercial Forestry Research (ICFR).

Lee, E., Lee, S.-T., Mun, H.-S., Chung, S. & Oh, J.-H. 2024. Assessing the residual stand damage after thinning with different levels of mechanization. *Forests,* 15, 794.

Leverkus, A. B., Lindenmayer, D. B., Thorn, S. & Gustafsson, L. 2018. Salvage logging in the world's forests: Interactions between natural disturbance and logging need recognition. *Global Ecology and Biogeography,* 27, 1140–1154.

Little, K., van den Berg, G. & Fuller, G. 2002. Coppicing potential of Eucalyptus nitens: Results from a field survey. *Southern African Forestry Journal,* 2002, 31–38.

Little, K., van Staden, J. & Thomas, T. 2005. Effects of vegetation control on Eucalyptus grandis x E. camaldulensis volume and economics. *South African Journal of Botany,* 71, 418–425.

Little, K. M., Ahtikoski, A. & Morris, A. R. 2018. Rotation-end financial performance of vegetation control on Eucalyptus smithii in South Africa. *Southern Forests: A Journal of Forest Science,* 80, 241–250.

Lopes, E. S., Oliveira, F. M. & Droog, A. 2018. Damage to residual trees following commercial thinning by harvester and forwarder in a Pinus taeda stand in Southern Brazil. *Scientia Forestalis*, 46 (117), 1–12.

Macdicken, K. G., Sola, P., Hall, J. E., Sabogal, C., Tadoum, M. & DE Wasseige, C. 2015. Global progress toward sustainable forest management. *Forest Ecology and Management*, 352, 47–56.

Marchi, E., Chung, W., Visser, R., Abbas, D., Nordfjell, T., Mederski, P. S., Mcewan, A., Brink, M. & Laschi, A. 2018. Sustainable Forest Operations (SFO): A new paradigm in a changing world and climate. *Science of the Total Environment*, 634, 1385–1397.

Marshall, V. 2000. Impacts of forest harvesting on biological processes in northern forest soils. *Forest Ecology and Management*, 133, 43–60.

Mcewan, A., Brink, M. & Van Zyl, S. 2013. *Guidelines for Difficult Terrain Ground based Harvesting Operations in South Africa*. Pietermaritzburg: Institute for Commercial Forestry Research.

Mcewan, A., Magagnotti, N. & Spinelli, R. 2016. The effects of number of stems per stool on cutting productivity in coppiced Eucalyptus plantations. *Silva Fennica*, 50, 1–14.

Mendonça, N. D. P., Gomes, V. D. S., Ferreira, M. A., Cruz, B. R. & Gomide, L. R. 2022. Silvicultural tasks scheduling optimization: A case study of functions and methods. *Revista Árvore*, 46, e4602.

Morris, A. R. 2022. Changing use of species and hybrids in South African forest plantations. *Southern Forests: A Journal of Forest Science*, 84, 193–205.

Ndlovu, N. N., Little, K. M., Titshall, L. & Rolando, C. A. 2019. The impact of slash management, fertilisation and vegetation management on Pinus elliottii pulpwood growth and rotation-end yield. *South African Journal of Plant and Soil*, 36, 249–259.

Nieto Lawrence, J. S. 2022. *The Effects of Forest Residue Management on Topsoil Nutrient Dynamics, and on Early Growth Responses of Eucalyptus Grandis x Urophylla under Varied Levels of Fertilization*. Stellenbosch: Stellenbosch University.

Nyland, R. D. 2016. *Silviculture: Concepts and Applications*. Long Grove, IL: Waveland Press.

Oberholzer, F. 2021. The South African forestry and forest products industry 2019. *Annual Stats of SA Forestry Industry*. Pietermaritzburg: Forestry South Africa, 1–160.

Pawson, S. M., Brin, A., Brockerhoff, E. G., Lamb, D., Payn, T. W., Paquette, A. & Parrotta, J. 2013. Plantation forests, climate change and biodiversity. *Biodiversity and Conservation*, 22, 1203–1227.

Peerbhay, K., Germishuizen, I. & Ismail, R. 2019. Can remote sensing detect, monitor and track baboon-damaged Pinus plantations located in South Africa? *Journal of Spatial Science*, 64, 535–549.

Pereira da Silva Correia, T., Caldas Lopes, A. G., Faggion, F., Rodrigues de Brito, G. & Gomes de Sousa, S. F. 2020. Operational performance and costs of mechanized stump extraction sets. *Engenharia na Agricultura*, 28, 1–28.

Perera, P. & Vlosky, R. P. 2006. *A History of Forest Certification*. Baton Rouge: Louisiana Forest Products Development Center, School of Renewable Natural Resources.

Petrie, B. 2014. *South Africa: A Case for Biomass*. London: International Institute for Environment and Development.

Picchio, R., Mederski, P. S. & Tavankar, F. 2020. How and how much, do harvesting activities affect forest soil, regeneration and stands? *Current Forestry Reports*, 6, 115–128.

Ramantswana, M., Brink, M., Little, K., Spinelli, R. & Chirwa, P. 2021. Perspectives and drivers of modernisation of silviculture re-establishment in South Africa. *Southern Forests: A Journal of Forest Science,* 83, 79–87.

Ramantswana, M., Mcewan, A. & Spinelli, R. 2017. The effect of coppice management on stump volume recovery in mechanized operations. *Annals of Forest Science,* 74, 1–8.

Ramantswana, M., Mcewan, A. & Steenkamp, J. 2013. A comparison between excavator-based harvester productivity in coppiced and planted Eucalyptus grandis compartments in KwaZulu-Natal, South Africa. *Southern Forests: A Journal of Forest Science,* 75, 239–246.

Rametsteiner, E. & Simula, M. 2003. Forest certification—An instrument to promote sustainable forest management? *Journal of Environmental Management,* 67, 87–98.

Rawat, Y. S., Singh, G. & Tekleyohannes, A. T. 2024. Optimizing the benefits of invasive alien plants biomass in South Africa. *Sustainability,* 16, 1876.

Rietz, D. N. 2010. *The effects of compaction and residue management on soil properties and growth of Eucalyptus grandis at two sites in KwaZulu-Natal, South Africa.* Doctoral dissertation, University of KwaZulu-Natal, Pietermaritzburg, South Africa.

Sarkar, A. U. 1998. Sustainability, sustainable development and forest resources. *The International Journal of Sustainable Development & World Ecology,* 5, 164–171.

Schoenholtz, S. H., Van Miegroet, H. & Burger, J. 2000. A review of chemical and physical properties as indicators of forest soil quality: challenges and opportunities. *Forest Ecology and Management,* 138, 335–356.

Schönau, A. 1989. Requirements for intensive silviculture. *South African Forestry Journal,* 150, 40–49.

Schwegman, K., Little, K.M., McEwan, A. and Ackerman, S.A., 2018. Harvesting and extraction impacts on Eucalyptus grandis× E. urophylla coppicing potential and rotation-end volume in Zululand, South Africa. *Southern Forests: A Journal of Forest Science,* 80 (1), 51–57.

Schweier, J., Magagnotti, N., Labelle, E. R. & Athanassiadis, D. 2019. Sustainability impact assessment of forest operations: A review. *Current Forestry Reports,* 5, 101–113.

Silva, N. F. D., Barros, N. F. D., Neves, J. C. L., Schulthais, F., Novais, R. F. D. & Mattiello, E. M. 2020. Yield and nutrient demand and efficiency of eucalyptus under coppicing regime. *Forests,* 11, 852.

Siry, J. P., Cubbage, F. W. & Ahmed, M. R. 2005. Sustainable forest management: Global trends and opportunities. *Forest Policy and Economics,* 7, 551–561.

Smith, C., Johnston, M. & Lorentz, S. 1997. Assessing the compaction susceptibility of South African forestry soils. I. The effect of soil type, water content and applied pressure on uni-axial compaction. *Soil and Tillage Research,* 41, 53–73.

Smith, C. W. 2000. Harvesting impacts and long-term site productivity: The South African experience. *In:* Dyck, W. J., Cole, D. W. & Comerford, N. B. (eds.) *Impacts of Forest Harvesting on Long-Term Site Productivity*. Helsinki: Metsäntutkimuslaitos, 14–22. ISBN: 951-40-1778-1.

Smith, C. W. & du Toit, B. 2005. The effect of harvesting operations, slash management and fertilisation on the growth of a Eucalyptus clonal hybrid on a sandy soil in Zululand, South Africa. *Southern African Forestry Journal,* 2005, 15–26.

Snowdon, P. 2002. Modeling Type 1 and Type 2 growth responses in plantations after application of fertilizer or other silvicultural treatments. *Forest Ecology and Management,* 163, 229–244.

Spinelli, R., Magagnotti, N. & Schweier, J. 2017. Trends and perspectives in coppice harvesting. *Croatian Journal of Forest Engineering: Journal for Theory and Application of Forestry Engineering,* 38, 219–230.

Tavankar, F., Nikooy, M., Lo Monaco, A., Latterini, F., Venanzi, R. & Picchio, R. 2020. Short-term recovery of residual tree damage during successive thinning operations. *Forests,* 11, 731.

Titshall, L., Dovey, S. & Rietz, D. 2013. A review of management impacts on the soil productivity of South African commercial forestry plantations and the implications for multiple-rotation productivity. *Southern Forests: A Journal of Forest Science,* 75, 169–183.

Titus, B. D., Brown, K., Helmisaari, H.-S., Vanguelova, E., Stupak, I., Evans, A., Clarke, N., Guidi, C., Bruckman, V. J. & Varnagiryte-Kabasinskiene, I. 2021. Sustainable forest biomass: A review of current residue harvesting guidelines. *Energy, Sustainability and Society,* 11, 1–32.

Udali, A., Chung, W., Talbot, B. & Grigolato, S. 2025. Managing harvesting residues: A systematic review of management treatments around the world. *Forestry: An International Journal of Forest Research,* 98, 117–135.

van der Ven, H. & Cashore, B. 2018. Forest certification: The challenge of measuring impacts. *Current Opinion in Environmental Sustainability,* 32, 104–111.

Van Wyk, G. 1978. Does tree improvement influence the harvesting and transport of trees? *South African Forestry Journal,* 105, 13–17.

Walmsley, J. & Godbold, D. 2010. Stump harvesting for bioenergy–A review of the environmental impacts. *Forestry,* 83, 17–38.

Walmsley, J., Jones, D., Reynolds, B., Price, M. & Healey, J. 2009. Whole tree harvesting can reduce second rotation forest productivity. *Forest Ecology and Management,* 257, 1104–1111.

Wingfield, M., Slippers, B., Hurley, B., Coutinho, T., Wingfield, B. & Roux, J. 2008. Eucalypt pests and diseases: Growing threats to plantation productivity. *Southern Forests: A Journal of Forest Science,* 70, 139–144.

Worrell, R. & Hampson, A. 1997. The influence of some forest operations on the sustainable management of forest soils—A review. *Forestry: An International Journal of Forest Research,* 70, 61–85.

Xulu, S., Gebreslasie, M. T. & Peerbhay, K. Y. 2019. Remote sensing of forest health and vitality: A South African perspective. *Southern Forests: A Journal of Forest Science,* 81, 91–102.

Zitha, H. M., Ramantswana, M. M. & Spinelli, R. 2023. Performance of harvesting residue treatment methods in South African pine plantations. *Forests,* 14, 1661.

Zobel, B. 1992. Silvicultural effects on wood properties. *IPEF International,* 2, 31–38.

7 Conclusion

Muedanyi M. Ramantswana

7.1 Introduction

Silviculture operations are critical for establishing and maintaining sustainably managed plantation forests. Silviculture practices differ worldwide because of regional preferences, terrain conditions, physiography, climate, labour availability and capacity, tree properties, and management regimes. The South African and Brazilian forestry industries need to innovate and adopt technologies to help reduce wood production costs and enhance their competitiveness in the global market. Some silviculture technologies may work efficiently in other parts of the world but may not necessarily be suited for certain conditions. Over the past two decades, the forestry industry has seen the introduction and acceleration of modern silviculture technologies in South Africa and Brazil. The adoption of these technologies is evident in nurseries, re-establishment efforts, and stand-tending operations.

Operational methods vary from manual to motor manual, to semi-mechanised, and ultimately to fully mechanised systems. The operational evidence is evident in the application of mulchers versus burning, pitting machines versus picks, tractor-based planting systems versus manual planting, and drones versus knapsacks. This demonstrates the progression in technology from manual to mechanised systems. This wave has been driven by the need to increase productivity while reducing costs, enhance quality (survival, tree growth, and uniformity), and improve safety. However, despite the gradual increase in modernisation within our industry, most of our silviculture activities are still performed manually; the industry is, nonetheless, being increasingly and rapidly transformed by innovative technologies.

7.2 Aim and scope of the book

The objective of this book was to comprehensively describe the various technologies used in silviculture, with a special focus on newer technologies adopted over the past two decades in plantation forestry practised in South Africa and Brazil. This book primarily focused on re-establishment, tending, and, to a lesser extent, harvesting technologies used in the two countries.

DOI: 10.4324/9781003391227-7

To recap, Chapter 1 provided an overview of plantation forestry practices, technology, and planning involved in the establishment/re-establishment and tending of a forest. Chapter 2 focused on nurseries, describing the resources and technologies used in plant production. Chapter 3 discussed the methods and innovations employed in regenerating/re-establishing forests, focusing on activities such as residue management, preparing planting positions, weeding before planting, tree planting, and fertilisation. Chapter 4 described stand-tending activities and the latest tools and resources available for executing activities such as weeding after planting, forest protection, pruning, and thinning. Chapter 5 provided a reflection on past, emerging, and future trends in the adoption of silviculture technology. Chapter 6 dealt with the relationship between silviculture and harvesting operations, specifically the impacts of establishment and tending activities on harvesting operations, as well as the effects of harvesting operations on silviculture operations and stand characteristics.

7.3 Key concepts

Modernisation of silviculture has increased over the past two decades. This is evident in the changes in the manner in which silvicultural tasks are performed, the introduction of new tools and equipment, the increased use of innovative resources (e.g. chemicals), and the utilisation of data. The adoption and rate of technology adoption is being driven by the need to improve health and safety, increase operational productivity and reduce costs of operations, and improve stand productivity (quality), social aspects/issues (mainly labour-related), environmental aspects/issues, and other aspects/issues, such as climate change and global competition. However, technology adoption has been hindered by several factors, including the lack of available technology, insufficient investment in technology development, the complexity of some silvicultural tasks, physical-environmental barriers, socio-economic barriers, and psychological factors. The key concepts from each chapter will be described in the section below.

7.3.1 Nursery technologies

The seedling or plant is a critical resource for the establishment of a forest. Without a good quality plant to grow, any subsequent activities performed will be in vain. Hence, it is essential that high-quality plants are produced; this process is realised at the nursery facility, which is purposefully designed to facilitate the establishment and growth of the plants. The key concepts to grasp regarding nursery technologies are that plants in South Africa and Brazil are predominantly produced using containerised systems. The process starts with the harvesting of cuttings and container filling with growing media, then plants are taken through the stages of plant development (germination/rooting, growing, hardening off, and dispatch). Throughout the different phases, watering and fertigation processes can be applied using precision technologies to moderate application rates. The continual assessment of plant quality traits, including morphological and physiological attributes, ensures that the plants produced meet the required field specifications. This assessment is even more critical when checking plants prepared for mechanised planting, which is deemed to be a harsher environment for fragile plants.

7.3.2 Establishment/re-establishment technologies

The establishment/re-establishment of a forest stand is a critical phase with direct implications for the yield generated at the end of the rotation. The species produced at the nursery need to be matched to the site where they are going to be planted. However, before the plants can be established, the environment needs to be prepared to facilitate the survival, growth, and uniformity of the trees. The establishment and re-establishment phase comprises the following activities: residue management, preparation of planting positions, weeding, tree planting, and fertilisation. These activities are performed using technologies that range from manual techniques to mechanised. However, some advanced technologies are still in the exploratory phase due to limitations such as lower productivity and higher costs.

The key concepts outlined were that harvest residues can hinder establishment activities (e.g. soil preparation), but there are various methods for managing residues, including burning, spreading, mulching, and biomass removal. Thereafter, marking a planting position to achieve the required stand spacing was described. Methods included manual marking with ropes or wires, and the use of GPS technology to identify areas where soil must be prepared. Once the appropriate position has been identified, the preparation of a planting position can be executed. The breaking of the soil to facilitate tree growth can be achieved through pitting or subsoiling. This can be achieved by using manual, motor manual, and mechanised soil preparation methods.

Weeding before planting was identified as a critical silvicultural practice. Methods included manual chemical spraying, tractor boom spraying, and the use of unmanned aerial vehicles (UAVs). Planting methods included manual planting, planting with planting tubes, semi-mechanised planting, and mechanised planting. Furthermore, planting can be executed with water or hydrogels (irrigation) to aid in survival and early growth. Fertilising is an essential activity in intensive plantation forestry, with manual and mechanised methods discussed. Finally, regeneration through coppice, a method to re-establish eucalypts after clear-felling, was presented. All the activities described are part of the establishment value chain, and when one is not performed to the required standards, it can have an adverse effect on subsequent activities.

7.3.3 Stand-tending technologies

The stand-tending phase comprises management activities to optimise tree growth and timber value. During this phase, the trees have been established on the site, and the growth of the trees that survived needs to be enhanced by performing activities such as weeding after planting, pruning, protection from pests, diseases, and fires. By conducting thinning, the residual tree growth can be optimised to the landowners' objectives and market requirements.

The key concepts described in stand tending include weeding after planting activities and technologies that can be used to perform the weeding operation. Various weeding methods were used, ranging from manual to mechanised and chemical approaches. Manual weeding is more labour-intensive and has lower productivity, but it is flexible. In contrast, mechanised methods offer higher productivity but may have machine-specific limitations. Of all the weeding methods, chemical control, using herbicides, was a predominant method for prolonged weed suppression; however, non-chemical methods are preferred due to their positive effect on the environment and health.

Tree pruning is a tending activity that was described. Pruning facilitates the production of improved wood quality by cutting branches to produce knot-free timber and improved stand access. Manual and motor manual pruning methods were identified as the two available methods commonly used between the two countries, with very limited mechanised options available in the market. Forest protection is another critical stand-tending activity that can be focused on biotic (pests, diseases) and abiotic (fire) agents. If not proactively managed, significant losses of trees, financial, and non-financial resources can occur over a relatively short period of time. Pest and disease control can be achieved through a holistic approach that combines prevention, eradication, and management – utilising various surveillance technologies to monitor forest health (e.g. remote sensing). Wildfire protection employs a multi-faceted approach that combines detection, prevention, and suppression methods, utilising a range of technologies. Firefighting techniques can be manual, semi-mechanised, or mechanised, but the most effective approach is integrated fire management, which combines various environment-specific strategies with ecological considerations. The final key concept described in this chapter is thinning, which is a silvicultural practice that influences stand structure and growth. Felling, extraction, and processing technologies used in thinning can range from semi-mechanised to mechanised methods. It is important to recognise that thinning operations are more costly compared to clear-fell harvesting because of the smaller tree size and more moving time.

7.3.4 Technological innovations in silviculture

Technology is advancing quickly, with forestry increasingly integrating various innovations. Local and global drivers in both South Africa and Brazil are driving the expansion of technology in forestry operations. These drivers include health and safety improvements, increased productivity with cost reduction, enhanced stand productivity, as well as social challenges (mainly labour-related).

The key concepts discussed in this chapter were the emerging or prospective future technologies classified into four main categories: machine-specific technical innovations, material input innovations, operator-specific innovations, and computerised technology applications. Machine-specific technical innovations, such as multifunctional machines, machine terrain handling enhancements, and machine automation/robotics, are essential technologies to consider in addressing safety, productivity, and cost reduction needs. Material inputs innovations, such as the use of paper-based pots, nano-fertilisers, fertiliser tablets, and ultra-low emission engines, are relevant in addressing plant quality (e.g. survival in mechanised planting and plant growth) and climate change (reducing carbon emissions) needs. Operator-specific innovations, such as ergonomically friendly machine cabs and simulation training, can be effective measures to address health and safety, productivity, and social challenges by creating a human-friendly environment. Computerised technology applications, including artificial intelligence (AI), real-time monitoring of machine and operator performance, big data processing and application, and remote control of operating machines, are key technologies that will drive precision silviculture and address productivity and cost needs.

7.3.5 *Silviculture and harvesting relationship*

Even though a silvicultural system comprises three distinct parts – regeneration/establishment, tending, and harvesting – these stages are dependent on each other. This chapter makes a distinction between silviculture and harvesting. Within the context of this book, the term silviculture is used with a focus on the regeneration and tending process stages only. Silviculture activities have an impact on harvesting activities and vice versa. Actions taken at the beginning of the value chain can affect other processes at the end of the value chain. The management of impacts is vital to creating and maintaining sustainable forestry practices. At the foundation of sustainable forestry practices is planning, which ensures that the economic, environmental, and social interests of the business and society are relatively balanced.

The key concepts discussed in this chapter were harvesting effects on modern silviculture. The impacts of each activity were described, along with intervention measures. The harvesting effects discussed are site impacts, biomass recovery, high stumps, coppice regeneration capacity, and delays in timber harvesting and extraction. In summary, the effects of these activities can result in poor survival, lesser tree growth, variable stand uniformity, limited stand access, and high silvicultural costs.

The effect of silviculture operations on harvesting was also discussed. The silviculture effects outlined were the execution of regeneration activities, planting line orientation, coppicing, vegetation management, tree morphology, thinning, and tree protection. In summary, the silviculture effects can result in low harvesting productivity due to smaller tree sizes and obstacles (e.g. vegetation obscuring the operators' view), higher safety risks, increased waste, and low timber recovery due to tree damage.

7.4 Challenges and opportunities

Some of the key challenges to silviculture modernisation within the South African context are the high cost of capital, potential reduction in manually orientated employment opportunities, low utilisation of equipment due to the seasonality of operations, inadequate economies of scale (especially for small landowners), and insufficient information on the costs and benefits of adopting new technologies. The scenario in Brazil differs significantly due to its large plantation forest expansion rates, increased investments, and the presence of large corporate forestry companies; consequently, most constraints faced in South Africa are less restrictive.

On the ground, the physical-environmental factors have influenced technology adoption in both South Africa and Brazil. Physical-environmental barriers refer to the prevalence of site-impeding factors such as terrain conditions (slope, ground conditions, and roughness), stumps, and harvesting residues. These factors predominantly affect accessibility and trafficability on the sites to execute silvicultural activities. Given the general lack of purpose-built equipment, forestry organisations resort to manual methods, which are more flexible. Socio-economic factors refer to the impact of modernisation on employment among rural communities that depend on forestry. South Africa must balance the need to improve efficiencies with the preservation and improvement of working conditions for people dependent on forestry work. However, in Brazil, some plantation forests are in isolated rural areas with limited labour availability, necessitating the use of advanced tools and machines to perform silvicultural tasks.

By using precision forestry technologies such as remote sensing and drones for monitoring and data collection, silviculture managers can be well informed about the status of their plantations. Precision systems can provide detailed information about climatic conditions (long-term weather patterns), forest stands (e.g. fuel loads, weed status, stress), and work performed by machines (e.g. telematic systems) and people within the forest stands. This information is valuable because managers can plan accurately and make informed operational decisions. Accurate information combined with good management can reduce costs and improve efficiencies. For example, during the establishment phase, South African forest grower companies can use UAV technologies to collect high-resolution spatial data to assess seedling survival rates, weeds, and seedling growth. Managers should configure their precision support systems to deliver relevant information that facilitates impactful decision-making.

The Brazilian forestry industry is almost ten times the size of the South African forestry industry. This scale offers opportunities for investing in technology on a large scale. This presents an opportunity for technology development, as machine manufacturers can invest resources in the market due to the existing and growing demand. Opportunities in South Africa are limited in this area due to the country's limited forest land area and complex landownership dynamics. However, there is an opportunity for both semi-mechanised and mechanised solutions that are fit for purpose. Equipment with relatively low capital costs, viable productivity, and good quality outputs can be a consideration for this market.

7.5 Final considerations

The future of a sustainable forestry industry comprises a diverse mix of knowledge, skills, and expertise. To fully realise the value of modern silviculture, it is essential to understand both the silvicultural principles of growing trees and the technologies that can be integrated to optimise manpower, machines, and materials. Forestry industries need to make a concerted effort to develop resourceful solutions collaboratively. Grower companies, contractors, government institutions, machine manufacturers, and educational institutions can be more organised and focused on innovating and studying the solutions most fitting for industry needs instead of waiting for remedies from single entities or abroad. Local collaborative entities can yield more impactful outputs if we break down self-made boundaries, find synergy, and see the bigger picture.

Glossary

Artificial intelligence (AI) is a wide-ranging branch of computer technologies concerned with building intelligent machines capable of augmenting, automating, and accelerating key day-to-day tasks associated with human intelligence

Automatic operation any activity using tools controlled by various artificial systems, with a high degree of technology and low levels of human intervention

Big data includes the use of cloud computing, Internet of Things (IoT), and mobile internet

Biomass any organic, non-fossil material from plants that can be used as a source of energy or for other products (e.g. forest residues)

Burnt forest stands forest sites where logging residue has been cleared through controlled burning

Broadcasting spreading evenly of the forest residue after a stand harvest

Chemical weed control the application of herbicides on weeds or pesticides on trees

Chopper rolling is a residue management process in which a heavy drum with parallel metal blades is pulled over residues on a site to break and reduce the size of the material retained on the site

Clear-fell system managing successive even-aged stands by clear-felling the entire stand of trees at planned rotations and then re-establishing and tending a new stand in place of the original one

Compaction the compression of soil by heavy machinery – skidding or on depots – causing permanent damage in terms of poor growth

Container seedling a seedling grown in a small container in a controlled environment

Coppicing process of allowing regeneration from coppice shoots for an additional one after clear-felling

Cuttings asexual propagation of plants using portions of young shoots of the mother tree

Disc harrowing is the process where a disc plough is used to uproot, cover, and damage weeds

Fertilisation the addition of fertiliser to promote tree growth on sites deficient in one or more soil nutrients. Also used to improve the vigour of crop trees following juvenile spacing or commercial thinning

Forest technology the study of the operations, work methods, logistics, and economics related to forest resources utilisation

Growing media a substrate or a soilless medium. Good media needs to provide physical support to the plant, adequate aeration, good water holding capacity, and supply mineral nutrients

Harvesting method the form in which trees are felled and extracted to the roadside for further processing

Harvesting residues woody materials such as leaves, twigs, branches, bark, treetops, stumps, and merchantable logs left in the field post-logging operations

Harvesting system are tools, equipment, and machines used to harvest an area

Integrated fire management combining the various components of fire management (fire prevention, detection, suppression, and use) with consideration of ecological systems

Irrigation the practice of applying controlled amounts of water to help trees grow, especially after planting

Machine terrain handling enhancements technologies to enable silviculture equipment to handle poor soil conditions, rougher ground conditions, and steeper slopes

Manual operation any activity using tools that are exclusively human-powered, e.g. a pick for pitting, a slasher for weeding, a planting tube for planting

Mechanical site preparation any activity that involves the use of mechanical machinery to prepare a site for reforestation

Mechanical weeding the use of equipment such as a tractor with an implement to cut or dig up and mix the weeds into the soil

Mechanised operation any activity using mechanical assemblies or implements that are exclusively mechanically powered with or without auto-pilot functions

Mulching is a mechanical treatment that converts plant residue from the previous harvest into mulch

Paper pots biodegradable containers for holding the growing media and facilitating root growth for seedlings

Planning the determination of the goals and objectives of an enterprise and the selection, through a systematic consideration of alternatives, of the policies, programmes, and procedures for achieving them. An activity devoted to clearly identifying, defining, and determining courses of action, before their initiation, that are necessary to achieve

Plantation forests intensively managed forests planted at regular spacing, even-aged, and composed of one or two species for the production of wood, energy fibre, and non-timber forest products

Planted forests include forests that have been established through planting or deliberate seeding

Planting involves carefully placing a plant into adequately prepared soil or establishing a forest by setting out seedlings, transplants, or cuttings in an area

Precision forestry the use of advanced machines, sensing, and analytical tools to gather accurate information to support decision-making in forest management

Preparation of a planting position breaking the soil after felling to facilitate the growth of a new generation of trees

Pruning is a wood quality improvement operation wherein branches from a tree stem are cut off to produce knot-free timber, also known as clear wood

Remote sensing the science of obtaining information about an object or phenomenon from a distance, typically using satellites or aircraft to detect and measure reflected or emitted energy

Residues material generated during the stand tending and harvesting of trees

Robot is an automatic device that executes tasks generally performed by a human

Semi-mechanised operation any activity that involves human labour with the assistance of a machine to perform silviculture operations

Silviculture is the art and science of cultivating forest crops through managing the establishment, growth, composition, health, and quality of forests and woodlands to meet the diverse needs of landowners and society sustainably

Silviculture planning a process of custom designing and optimising the density of regeneration, the timing and intensity of stand tending treatments, and the timing of future harvests

Silviculture system is a planned programme of silvicultural treatments throughout a stand's entire life, involving cutting, reproduction, tending operations, and intermediate cuttings

Silviculture technology can be described as the application of scientific knowledge to study modern tools, machinery, and equipment for establishing, tending, and growing trees

Stump clipping involves cutting stumps as close as possible to the ground without removing the entire stump (root system)

Subsoiling mechanical soil preparation method that opens a narrow trench either by loosening and or breaking hard subsoil layers using a metal tine (can be up to 1 m)

Thinning the selective removal of plants, trees, or parts of plants to encourage the growth of remaining ones

Tree morphology refers to the characteristics of a tree, including its stem, branchiness, and canopy structure

Unmanned aerial vehicle is a powered aerial vehicle that does not carry a human operator and uses aerodynamic forces to provide vehicle lift. It can fly autonomously or be piloted remotely, is expendable or recoverable, and can carry a payload

Wasserplanter a tractor towing a high-pressure water planting rig which combines soil preparation, irrigation, and planting in a single pass

Weeding removing unwanted vegetation (e.g. grass, herbs, shrubs, and coppice), which can suppress and affect the trees in the early growth stage

Windrowing is a site preparation process in which surface residue material is moved and accumulated in the form of a windrow (long parallel lines) manually or mechanically

Index

Note: **Bold** page numbers refer to tables and *italic* page numbers refer to figures.